D0230138

Tynnwyd yn ôl
SWANSEA LIBRARIES
Withdrawn
6000340697

The
Uninhabitable
Earth

The
Uninhabitable
Earth

A Story of the Future

David Wallace-Wells

ALLEN LANE
an imprint of
PENGUIN BOOKS

ALLEN LANE

UK | USA | Canada | Ireland | Australia
India | New Zealand | South Africa

Penguin Books is part of the Penguin Random House group of companies
whose addresses can be found at global.penguinrandomhouse.com.

Penguin
Random House
UK

First published in the United States of America by Tim Duggan Books,
an imprint of the Crown Publishing Group,
a division of Penguin Random House LLC, 2019
First published in Great Britain by Allen Lane 2019
001

Copyright © David Wallace-Wells, 2019

The moral right of the author has been asserted

Set in Adobe Caslon Pro 11.82/16.70 pt by Jouve (UK), Milton Keynes
Printed and bound in Great Britain by Clays Ltd, Elcograf S.p.A.

A CIP catalogue record for this book is available from the British Library

ISBN: 978–0–241–35521–3

CITY AND COUNTY OF SWANSEA LIBRARIES	
6000340697	
Askews & Holts	05-Mar-2019
363.738	£20.00
SWKL	

www.greenpenguin.co.uk

MIX
Paper from
responsible sources
FSC   FSC® C018179
www.fsc.org

Penguin Random House is committed to a
sustainable future for our business, our readers
and our planet. This book is made from Forest
Stewardship Council® certified paper.

For Risa and Rocca,
My mother and father

Contents

The
Uninhabitable
Earth

I

Cascades

I T IS WORSE, MUCH WORSE, THAN YOU THINK. THE SLOWNESS of climate change is a fairy tale, perhaps as pernicious as the one that says it isn't happening at all, and comes to us bundled with several others in an anthology of comforting delusions: that global warming is an Arctic saga, unfolding remotely; that it is strictly a matter of sea level and coastlines, not an enveloping crisis sparing no place and leaving no life undeformed; that it is a crisis of the "natural" world, not the human one; that those two are distinct, and that we live today somehow outside or beyond or at the very least defended against nature, not inescapably within and literally overwhelmed by it; that wealth can be a shield against the ravages of warming; that the burning of fossil fuels is the price of continued economic growth; that growth, and the technology it produces, will allow us to engineer our way out of environmental disaster; that there is any analogue to the scale or scope of this threat, in the long span of human history, that might give us confidence in staring it down.

None of this is true. But let's begin with the speed of change. The earth has experienced five mass extinctions before the one we are living through now, each so complete a wiping of the fossil record that it functioned as an evolutionary reset, the planet's phylogenetic tree first expanding, then collapsing, at intervals, like a lung: 86 percent of all species dead, 450 million years ago; 70 million years later, 75 percent; 125 million years later, 96 percent; 50 million years later, 80 percent; 135 million years after that, 75 percent again. Unless you are a teenager, you probably read in your high school textbooks that these extinctions were the result of asteroids. In fact, all but the one that killed the dinosaurs involved climate change produced by greenhouse gas. The most notorious was 250 million years ago; it began when carbon dioxide warmed the planet by five degrees Celsius, accelerated when that warming triggered the release of methane, another greenhouse gas, and ended with all but a sliver of life on Earth dead. We are currently adding carbon to the atmosphere at a considerably faster rate; by most estimates, at least ten times faster. The rate is one hundred times faster than at any point in human history before the beginning of industrialization. And there is already, right now, fully a third more carbon in the atmosphere than at any point in the last 800,000 years—perhaps in as long as 15 million years. There were no humans then. The oceans were more than a hundred feet higher.

Many perceive global warming as a sort of moral and economic debt, accumulated since the beginning of the Industrial Revolution and now come due after several centuries. In fact, more than half of the carbon exhaled into the atmosphere by the burning of fossil fuels has been emitted in just the past three decades. Which means we have done as much damage to the fate of the planet and its ability to sustain human life and civilization since Al Gore published his first book on climate than in all the centuries—all the millennia—that came before. The United Nations established its climate change framework in 1992, advertising scientific consensus unmistakably to the world; this means we have now engineered as much ruin knowingly as we ever managed in ignorance. Global warming may seem like a distended morality tale playing out over several centuries and inflicting a kind of Old Testament retribution on the great-great-grandchildren of those responsible, since it was carbon burning in eighteenth-century England that lit the fuse of everything that has followed. But that is a fable about historical villainy that acquits those of us alive today—and unfairly. The majority of the burning has come since the premiere of *Seinfeld*. Since the end of World War II, the figure is about 85 percent. The story of the industrial world's kamikaze mission is the story of a single lifetime—the planet brought from seeming stability to the brink of catastrophe in the years between a baptism or bar mitzvah and a funeral.

We all know those lifetimes. When my father was born in 1938—among his first memories the news of Pearl Harbor and the

mythic air force of the industrial propaganda films that followed—
the climate system appeared, to most human observers, steady.
Scientists had understood the greenhouse effect, had understood
the way carbon produced by burned wood and coal and oil could
hothouse the planet and disequilibrate everything on it, for three-
quarters of a century. But they had not yet seen the impact, not
really, not yet, which made warming seem less like an observed fact
than a dark prophecy, to be fulfilled only in a very distant future—
perhaps never. By the time my father died, in 2016, weeks after
the desperate signing of the Paris Agreement, the climate system
was tipping toward devastation, passing the threshold of carbon
concentration—400 parts per million in the earth's atmosphere,
in the eerily banal language of climatology—that had been, for
years, the bright red line environmental scientists had drawn in the
rampaging face of modern industry, saying, *Do not cross*. Of course,
we kept going: just two years later, we hit a monthly average of 411,
and guilt saturates the planet's air as much as carbon, though we
choose to believe we do not breathe it.

The single lifetime is also the lifetime of my mother: born in
1945, to German Jews fleeing the smokestacks through which their
relatives were incinerated, and now enjoying her seventy-third year
in an American commodity paradise, a paradise supported by the
factories of a developing world that has, in the space of a single lifetime, too, manufactured its way into the global middle class, with
all the consumer enticements and fossil fuel privileges that come
with that ascent: electricity, private cars, air travel, red meat. She
has been smoking for fifty-eight of those years, unfiltered, ordering
the cigarettes now by the carton from China.

It is also the lifetime of many of the scientists who first raised
public alarm about climate change, some of whom, incredibly, remain working—that is how rapidly we have arrived at this promontory. Roger Revelle, who first heralded the heating of the planet,
died in 1991, but Wallace Smith Broecker, who helped popularize
the term "global warming," still drives to work at the Lamont-
Doherty Earth Observatory across the Hudson every day from the

Upper West Side, sometimes picking up lunch at an old Jersey fill-
ing station recently outfitted as a hipster eatery; in the 1970s, he
did his research with funding from Exxon, a company now the
target of a raft of lawsuits that aim to adjudicate responsibility for
the rolling emissions regime that today, barring a change of course
on fossil fuels, threatens to make parts of the planet more or less
unlivable for humans by the end of this century. That is the course
we are speeding so blithely along—to more than four degrees Celsius of warming by the year 2100. According to some estimates,
that would mean that whole regions of Africa and Australia and
the United States, parts of South America north of Patagonia, and
Asia south of Siberia would be rendered uninhabitable by direct
heat, desertification, and flooding. Certainly it would make them
inhospitable, and many more regions besides. This is our itinerary,
our baseline. Which means that, if the planet was brought to the
brink of climate catastrophe within the lifetime of a single generation, the responsibility to avoid it belongs with a single generation,
too. We all also know that second lifetime. It is ours.

I AM NOT AN ENVIRONMENTALIST, AND DON'T EVEN THINK OF
myself as a nature person. I've lived my whole life in cities, enjoying gadgets built by industrial supply chains I hardly think twice
about. I've never gone camping, not willingly anyway, and while I
always thought it was basically a good idea to keep streams clean
and air clear, I also always accepted the proposition that there was
a trade-off between economic growth and cost to nature—and figured, well, in most cases I'd probably go for growth. I'm not about
to personally slaughter a cow to eat a hamburger, but I'm also not
about to go vegan. I tend to think when you're at the top of the food
chain it's okay to flaunt it, because I don't see anything complicated
about drawing a moral boundary between us and other animals,
and in fact find it offensive to women and people of color that all of
a sudden there's talk of extending human-rights-like legal protections to chimps, apes, and octopuses, just a generation or two after

we finally broke the white-male monopoly on legal personhood. In these ways—many of them, at least—I am like every other American who has spent their life fatally complacent, and willfully deluded, about climate change, which is not just the biggest threat human life on the planet has ever faced but a threat of an entirely different category and scale. That is, the scale of human life itself.

A few years ago, I began collecting stories of climate change, many of them terrifying, gripping, uncanny narratives, with even the most small-scale sagas playing like fables: a group of Arctic scientists trapped when melting ice isolated their research center, on an island populated also by a group of polar bears; a Russian boy killed by anthrax released from a thawing reindeer carcass, which had been trapped in permafrost for many decades. At first, it seemed the news was inventing a new genre of allegory. But of course climate change is not an allegory.

Beginning in 2011, about one million Syrian refugees were unleashed on Europe by a civil war inflamed by climate change and drought—and in a very real sense, much of the "populist moment" the entire West is passing through now is the result of panic produced by the shock of those migrants. The likely flooding of Bangladesh threatens to create ten times as many, or more, received by a world that will be even further destabilized by climate chaos—and, one suspects, less receptive the browner those in need. And then there will be the refugees from sub-Saharan Africa, Latin America, and the rest of South Asia—140 million by 2050, the World Bank estimates, meaning more than a hundred times Europe's Syrian "crisis."

The U.N. projections are bleaker: 200 million climate refugees by 2050. Two hundred million was the entire world population at the peak of the Roman Empire, if you can imagine every single person alive and living anywhere on the planet at that time dispossessed of their home and turned outward to wander through hostile territories in search of a new one. The high end of what's possible in the next thirty years, the United Nations says, is considerably worse: "a billion or more vulnerable poor people with little choice but to fight or flee." A billion or more. That was the entire global population as recently as 1820, with the Industrial Revolution well under way. Which suggests that we might better conceive of history not as a deliberate procession of years marching forward on a timeline but as an expanding balloon of population growth, humanity dilating across the planet almost to the point of full eclipse. One reason carbon emissions have accelerated so much in the last generation is also an explanation for why history seems to be proceeding so much faster, with so much more happening, everywhere, each year, even every day: this is what results when there are simply that many more humans around. Fifteen percent of all human experience throughout history, it's been estimated, belongs to people alive right now, each walking the earth with carbon footprints.

Those refugee figures are high-end estimates, produced years ago by research groups designed to call attention to a particular cause or crusade; the true numbers will almost surely fall short of them, and scientists tend to trust projections in the tens of millions rather than the hundreds of millions. But that those bigger numbers are only the far upper reaches of what is possible should not lull us into complacency; when we dismiss the worst-case possibilities, it distorts our sense of likelier outcomes, which we then regard as extreme scenarios we needn't plan so conscientiously for. High-end estimates establish the boundaries of what's possible, between which we can better conceive of what is likely. And perhaps they will prove better guides even than that, considering the optimists have never, in the half century of climate anxiety we've already endured, been right.

My file of stories grew daily, but very few of the clips, even those drawn from new research published in the most pedigreed scientific journals, seemed to appear in the coverage about climate change the country watched on television and read in its newspapers. In those places, climate change was reported, of course, and even with some tinge of alarm. But the discussion of possible effects was misleadingly narrow, limited almost invariably to the matter of sea-level rise. Just as worrisome, the coverage was san-guine, all things considered. As recently as the 1997 signing of the landmark Kyoto Protocol, two degrees Celsius of global warming was considered the threshold of catastrophe: flooded cities, crippling droughts and heat waves, a planet battered daily by hurricanes and monsoons we used to call "natural disasters" but will soon normalize as simply "bad weather." More recently, the foreign minister of the Marshall Islands offered another name for that level of warming: "genocide."

There is almost no chance we will avoid that scenario. The Kyoto Protocol achieved, practically, nothing; in the twenty years since, despite all of our climate advocacy and legislation and progress on green energy, we have produced more emissions than in the twenty years before. In 2016, the Paris accords established two degrees as a global goal, and, to read our newspapers, that level of warming remains something like the scariest scenario it is responsible to consider; just a few years later, with no single industrial nation on track to meet its Paris commitments, two degrees looks more like a best-case outcome, at present hard to credit, with an entire bell curve of more horrific possibilities extending beyond it and yet shrouded, delicately, from public view.

For those telling stories about climate, such horrific possibilities—and the fact that we had squandered our chance of landing anywhere on the better half of that curve—had become somehow unseemly to consider. The reasons are almost too many to count, and so half-formed they might better be called impulses. We chose not to discuss a world warmed beyond two degrees out of decency, perhaps; or simple fear; or fear of fearmongering; or technocratic faith, which is really market faith; or deference to partisan debates or even partisan priorities; or skepticism about the environmental Left of the kind I'd always had; or disinterest in the fates of distant ecosystems like I'd also always had. We felt confusion about the science and its many technical terms and hard-to-parse numbers, or at least an intuition that others would be easily confused about the science and its many technical terms and hard-to-parse numbers.

We suffered from slowness apprehending the speed of change, or semi-conspiratorial confidence in the responsibility of global elites and their institutions, or obeisance toward those elites and their institutions, whatever we thought of them. Perhaps we felt unable to really trust scarier projections because we'd only just heard about warming, we thought, and things couldn't possibly have gotten that much worse just since the first *Inconvenient Truth;* or because we liked driving our cars and eating our beef and living as we did in every other way and didn't want to think too hard about that; or because we felt so "postindustrial" we couldn't believe we were still drawing material breaths from fossil fuel furnaces. Perhaps it was because we were so sociopathically good at collating bad news into a sickening evolving sense of what constituted "normal," or because we looked outside and things seemed still okay. Because we were bored with writing, or reading, the same story again and again, because climate was so global and therefore nontribal it suggested only the corniest politics, because we didn't yet appreciate how fully it would ravage our lives, and because, selfishly, we didn't mind destroying the planet for others living elsewhere on it or those not yet born who would inherit it from us, outraged. Because we had too much faith in the teleological shape of history and the arrow of human progress to countenance the idea that the arc of history would bend toward anything but environmental justice, too. Because when we were being really honest with ourselves we already thought of the world as a zero-sum resource competition and believed that whatever happened we were probably going to continue to be the victors, relatively speaking anyway, advantages of class being what they are and our own luck in the natalist lottery being what it was. Perhaps we were too panicked about our own jobs and industries to fret about the future of jobs and industry; or perhaps we were also really afraid of robots or were too busy looking at our new phones; or perhaps, however easy we found the apocalypse reflex in our culture and the path of panic in our politics, we truly had a good-news bias when it came to the big picture; or, really, who knows why—there are so many aspects to the climate kaleidoscope that transforms our intuitions about environmental devastation into an uncanny complacency that it can be hard to pull the whole picture of climate distortion into focus. But we simply wouldn't, or couldn't, or anyway didn't look squarely in the face of the science.

THIS IS NOT A BOOK ABOUT THE SCIENCE OF WARMING; IT IS about what warming means to the way we live on this planet. But what does that science say? It is complicated research, because it is built on two layers of uncertainty: what humans will do, mostly in terms of emitting greenhouse gases, and how the climate will respond, both through straightforward heating and a variety of more complicated, and sometimes contradictory, feedback loops. But even shaded by those uncertainty bars it is also very clear research, in fact terrifyingly clear. The United Nations' Intergovernmental Panel on Climate Change (IPCC) offers the gold-standard assessments of the state of the planet and the likely trajectory for climate change—gold-standard, in part, because it is conservative, integrating only new research that passes the threshold of inarguability. A new report is expected in 2022, but the most recent one says that if we take action on emissions soon, instituting immediately all of the commitments made in the Paris accords but nowhere yet actually implemented, we are likely to get about 3.2 degrees of warming, or about three times as much warming as the planet has seen since the beginning of industrialization—bringing the unthinkable collapse of the planet's ice sheets not just into the realm of the real but into the present. That would eventually flood not just Miami and Dhaka but Shanghai and Hong Kong and a hundred other cities around the world. The tipping point for that collapse is said to be around two degrees; according to several recent studies, even a rapid cessation of carbon emissions could bring us that amount of warming by the end of the century.

The assaults of climate change do not end at 2100 just because most modeling, by convention, sunsets at that point. This is why some studying global warming call the hundred years to follow the "century of hell." Climate change is fast, much faster than it seems we have the capacity to recognize and acknowledge; but it is also long, almost longer than we can truly imagine.

In reading about warming, you will often come across analogies from the planetary record: *the last time the planet was this much warmer,* the logic runs, *sea levels were here.* These conditions are not coincidences. The sea level was there largely because the planet was that much warmer, and the geologic record is the best model we have for understanding the very complicated climate system and gauging just how much damage will come from turning up the temperature by two or four or six degrees. Which is why it is especially concerning that recent research into the deep history of the planet suggests that our current climate models may be underestimating the amount of warming we are due for in 2100 by as much as half. In other words, temperatures could rise, ultimately, by as much as double what the IPCC predicts. Hit our Paris emissions targets and we may still get four degrees of warming, meaning a green Sahara and the planet's tropical forests transformed into fire-dominated savanna. The authors of one recent paper suggested the warming could be more dramatic still—slashing our emissions could still bring us to four or five degrees Celsius, a scenario they said would pose severe risks to the habitability of the entire planet. "Hothouse Earth," they called it.

Because these numbers are so small, we tend to trivialize the differences between them—one, two, four, five. Human experience and memory offer no good analogy for how we should think of those thresholds, but, as with world wars or recurrences of cancer, you don't want to see even one. At two degrees, the ice sheets will begin their collapse, 400 million more people will suffer from water scarcity, major cities in the equatorial band of the planet will become unlivable, and even in the northern latitudes heat waves will kill thousands each summer. There would be thirty-two times as many extreme heat waves in India, and each would last five times as long, exposing ninety-three times more people. This is our best-case scenario. At three degrees, southern Europe would be in permanent drought, and the average drought in Central America would last nineteen months longer and in the Caribbean twenty-one months longer. In northern Africa, the figure is sixty months longer—five years. The areas burned each year by wildfires would double in the Mediterranean and sextuple, or more, in the United States. At four degrees, there would be eight million more cases of dengue fever each year in Latin America alone and close to annual global food crises. There could be 9 percent more heat-related deaths. Damages from river flooding would grow thirtyfold in Bangladesh, twentyfold in India, and as much as sixtyfold in the United Kingdom. In certain places, six climate-driven natural disasters could strike simultaneously, and, globally, damages could pass $600 trillion—more than twice the wealth as exists in the world today. Conflict and warfare could double.

Even if we pull the planet up short of two degrees by 2100, we will be left with an atmosphere that contains 500 parts per million of carbon—perhaps more. The last time that was the case, sixteen million years ago, the planet was not two degrees warmer; it was somewhere between five and eight, giving the planet about 130 feet of sea-level rise, enough to draw a new American coastline as far west as I-95. Some of these processes take thousands of years to unfold, but they are also irreversible, and therefore effectively permanent. You might hope to simply reverse climate change; you can't. It will outrun all of us.

This is part of what makes climate change what the theorist Timothy Morton calls a "hyperobject"—a conceptual fact so large and complex that, like the internet, it can never be properly comprehended. There are many features of climate change—its size, its scope, its brutality—that, alone, satisfy this definition; together they might elevate it into a higher and more incomprehensible conceptual category yet. But time is perhaps the most mind-bending feature, the worst outcomes arriving so long from now that we reflexively discount their reality.

Yet those outcomes promise to mock us and our own sense

of the real in return. The ecological dramas we have unleashed through our land use and by burning fossil fuels—slowly for about a century and very rapidly for only a few decades—will play out over many millennia, in fact over a longer span of time than humans have even been around, performed in part by creatures and in environments we do not yet even know, ushered onto the world stage by the force of warming. And so, in a convenient cognitive bargain, we have chosen to consider climate change only as it will present itself this century. By 2100, the United Nations says, we are due for about 4.5 degrees of warming, following the path we are on today. That is, farther from the Paris track than the Paris track is from the two-degree threshold of catastrophe, which it more than doubles.

As Naomi Oreskes has noted, there are far too many uncertainties in our models to take their predictions as gospel. Just running those models many times, as Gernot Wagner and Martin Weitzman do in their book *Climate Shock,* yields an 11 percent chance we overshoot six degrees. Recent work by the Nobel laureate William Nordhaus suggests that better-than-anticipated economic growth means better than one-in-three odds that our emissions will exceed the U.N.'s worst-case "business as usual" scenario. In other words, a temperature rise of five degrees or possibly more.

The upper end of the probability curve put forward by the U.N. to estimate the end-of-the-century, business-as-usual scenario— the worst-case outcome of a worst-case emissions path—puts us at eight degrees. At that temperature, humans at the equator and in the tropics would not be able to move around outside without dying.

In that world, eight degrees warmer, direct heat effects would be the least of it: the oceans would eventually swell two hundred feet higher, flooding what are now two-thirds of the world's major cities; hardly any land on the planet would be capable of efficiently producing any of the food we now eat; forests would be roiled by rolling storms of fire, and coasts would be punished by more and more intense hurricanes; the suffocating hood of tropical disease would reach northward to enclose parts of what we now call the Arctic; probably about a third of the planet would be made unlivable by direct heat; and what are today literally unprecedented and intolerable droughts and heat waves would be the quotidian condition of whatever human life was able to endure.

We will, almost certainly, avoid eight degrees of warming; in fact, several recent papers have suggested the climate is actually less sensitive to emissions than we'd thought, and that even the upper bound of a business-as-usual path would bring us to about five degrees, with a likely destination around four. But five degrees is nearly as unthinkable as eight, and four degrees not much better: the world in a permanent food deficit, the Alps as arid as the Atlas Mountains.

Between that scenario and the world we live in now lies only the open question of human response. Some amount of further warming is already baked in, thanks to the protracted processes by which the planet adapts to greenhouse gas. But all of those paths projected from the present—to two degrees, to three, to four, five, or even eight—will be carved overwhelmingly by what we choose to do now. There is nothing stopping us from four degrees other than our own will to change course, which we have yet to display. Because the planet is as big as it is, and as ecologically diverse; because humans have proven themselves an adaptable species, and will likely continue to adapt to outmaneuver a lethal threat; and because the devastating effects of warming will soon become too extreme to ignore, or deny, if they haven't already; because of all that, it is unlikely that climate change will render the planet truly uninhabitable. But if we do nothing about carbon emissions, if the next thirty years of industrial activity trace the same arc upward as the last thirty years have, whole regions will become unlivable by any standard we have today as soon as the end of this century.

A few years ago, E. O. Wilson proposed a term, "Half-Earth," to help us think through how we might adapt to the pressures of a changing climate, letting nature run its rehabilitative course on half the planet and sequestering humanity in the remaining, habitable half of the world. The fraction may be smaller than that, possibly considerably, and not by choice; the subtitle of his book was *Our Planet's Fight for Life*. On longer timescales, the even-bleaker outcome is possible, too—the livable planet darkening as it approaches a human dusk.

It would take a spectacular coincidence of bad choices and bad luck to make that kind of zero earth possible within our lifetime. But the fact that we have brought that nightmare eventuality into play at all is perhaps the overwhelming cultural and historical fact of the modern era—what historians of the future will likely study about us, and what we'd have hoped the generations before ours would have had the foresight to focus on, too. Whatever we do to stop warming, and however aggressively we act to protect ourselves from its ravages, we will have pulled the devastation of human life on Earth into view—close enough that we can see clearly what it would look like and know, with some degree of precision, how it will punish our children and grandchildren. Close enough, in fact, that we are already beginning to feel its effects ourselves, when we do not turn away.

IT IS ALMOST HARD TO BELIEVE JUST HOW MUCH HAS HAPpened and how quickly. In the late summer of 2017, three major hurricanes arose in the Atlantic at once, proceeding at first along the same route as though they were battalions of an army on the march. Hurricane Harvey, when it struck Houston, delivered such epic rainfall it was described in some areas as a "500,000-year event"—meaning that we should expect that amount of rain to hit that area once every five hundred millennia.

Sophisticated consumers of environmental news have already learned how meaningless climate change has rendered such terms, which were meant to describe storms that had a 1-in-500,000 chance of striking in any given year. But the figures do help in this way: to remind us just how far global warming has already taken us from any natural-disaster benchmark our grandparents would have recognized. To dwell on the more common 500-year figure just for a moment, it would mean a storm that struck once during the entire history of the Roman Empire. Five hundred years ago, there were no English settlements across the Atlantic, so we are talking about a storm that should hit just once as Europeans arrived and established colonies, as colonists fought a revolution and Americans a civil war and two world wars, as their descendants established an empire of cotton on the backs of slaves, freed them, and then brutalized *their* descendants, industrialized and post-industrialized, triumphed in the Cold War, ushered in the "end of history," and witnessed, just a decade later, its dramatic return. One storm in all that time, is what the meteorological record has taught us to expect. Just one. Harvey was the third such flood to hit Houston since 2015. And the storm struck, in places, with an intensity that was supposed to be a thousand times rarer still.

That same season, an Atlantic hurricane hit Ireland, 45 million were flooded from their homes in South Asia, and unprecedented wildfires tilled much of California into ash. And then there was the new category of quotidian nightmare, climate change inventing the once-unimaginable category of obscure natural disasters—crises so large they would once have been inscribed in folklore for centuries today passing across our horizons ignored, overlooked, or forgotten. In 2016, a "thousand-year flood" drowned small-town Ellicott City, Maryland, to take but one example almost at random; it was followed, two years later, in the same small town, by another. One week that summer of 2018, dozens of places all over the world were hit with record heat waves, from Denver to Burlington to Ottawa; from Glasgow to Shannon to Belfast; from Tbilisi, in Georgia, and Yerevan, in Armenia, to whole swaths of southern Russia. The previous month, the daytime temperature of one city in Oman reached above 121 degrees Fahrenheit, and did not drop below 108 all night, and in Quebec, Canada, fifty-four died from the heat. That same week, one hundred major wildfires burned in the American West, including one in California that grew 4,000 acres in one day, and another, in Colorado, that produced a volcano-like 300-foot eruption of flames, swallowing an entire sub-division and inventing a new term, "fire tsunami," along the way. On the other side of the planet, biblical rains flooded Japan, where 1.2 million were evacuated from their homes. Later that summer, Typhoon Mangkhut forced the evacuation of 2.45 million from mainland China, the same week that Hurricane Florence struck the Carolinas, turning the port city of Wilmington briefly into an island and flooding large parts of the state with hog manure and coal ash. Along the way, the winds of Florence produced dozens of tornadoes across the region. The previous month, in India, the state of Kerala was hit with its worst floods in almost a hundred years. That October, a hurricane in the Pacific wiped Hawaii's East Island entirely off the map. And in November, which has traditionally marked the beginning of the rainy season in California, the state was hit instead with the deadliest fire in its history—the Camp Fire, which scorched several hundred square miles outside of Chico, kill-ing dozens and leaving many more missing in a place called, prover-bially, Paradise. The devastation was so complete, you could almost forget the Woolsey Fire, closer to Los Angeles, which burned at the same time and forced the sudden evacuation of 170,000.

It is tempting to look at these strings of disasters and think, *Climate change is here.* And one response to seeing things long pre-dicted actually come to pass is to feel that we have settled into a new era, with everything transformed. In fact, that is how Califor-nia governor Jerry Brown described the state of things in the midst of the state's wildfire disaster: "a new normal."

The truth is actually much scarier. That is, the end of normal; never normal again. We have already exited the state of environ-mental conditions that allowed the human animal to evolve in the first place, in an unsure and unplanned bet on just what that animal can endure. The climate system that raised us, and raised every-thing we now know as human culture and civilization, is now, like a parent, dead. And the climate system we have been observing for the last several years, the one that has battered the planet again and again, is not our bleak future in preview. It would be more precise to say that it is a product of our recent climate past, already passing behind us into a dustbin of environmental nostalgia. There is no longer any such thing as a "natural disaster," but not only will things get worse; technically speaking, they have already gotten worse. Even if, miraculously, humans immediately ceased emitting carbon, we'd still be due for some additional warming from just the stuff we've put into the air already. And of course, with global emissions still increasing, we're very far from zeroing out on carbon, and therefore very far from stalling climate change. The devastation we are now seeing all around us is a beyond-best-case scenario for the future of warming and all the climate disasters it will bring.

What that means is that we have not, at all, arrived at a new equilibrium. It is more like we've taken one step out on the plank off a pirate ship. Perhaps because of the exhausting false debate about whether climate change is "real," too many of us have developed a misleading impression that its effects are binary. But global warming is not "yes" or "no," nor is it "today's weather forever" or "doomsday tomorrow." It is a function that gets worse over time as long as we continue to produce greenhouse gas. And so the experience of life in a climate transformed by human activity is not just a matter of stepping from one stable ecosystem into another, somewhat worse one, no matter how degraded or destructive the transformed climate is. The effects will grow and build as the planet continues to warm: from 1 degree to 1.5 to almost certainly 2 degrees and beyond. The last few years of climate disasters may look like about as much as the planet can take. In fact, we are only just entering our brave new world, one that collapses below us as soon as we set foot on it.

Many of these new disasters arrived accompanied by debate about their cause—about how much of what they have done to us comes from what we have done to the planet. For those hoping to better understand precisely how a monstrous hurricane arises out of a placid ocean, these inquiries are worthwhile, but for all practical purposes the debate yields no real meaning or insight. A particular hurricane may owe 40 percent of its force to anthropogenic global warming, the evolving models might suggest, and a particular drought may be half again as bad as it might have been in the seventeenth century. But climate change is not a discrete clue we can find at the scene of a local crime—one hurricane, one heat wave, one famine, one war. Global warming isn't a perpetrator; it's a conspiracy. We all live within climate and within all the changes we have produced in it, which enclose us all and everything we do. If hurricanes of a certain force are now five times as likely as in the pre-Columbian Caribbean, it is parsimonious to the point of triviality to argue over whether this one or that one was "climate-caused." All hurricanes now unfold in the weather systems we have wrecked on their behalf, which is why there are more of them, and why they are stronger. The same is true for wildfires: this one or that one may be "caused" by a cookout or a downed power line, but each is burning faster, bigger, and longer because of global warming, which gives no reprieve to fire season. Climate change isn't something happening here or there but everywhere, and all at once. And unless we choose to halt it, it will never stop.

Over the past few decades, the term "Anthropocene" has climbed out of academic discourse and into the popular imagination—a name given to the geologic era we live in now, and a way to signal that it is a new era, defined on the wall chart of deep history by human intervention. One problem with the term is that it implies a conquest of nature, even echoing the biblical "dominion." But however sanguine you might be about the proposition that we have already ravaged the natural world, which we surely have, it is another thing entirely to consider the possibility that we have only provoked it, engineering first in ignorance and then in denial a climate system that will now go to war with us for many centuries, perhaps until it destroys us. That is what Wally Broecker, the avuncular oceanographer, means when he calls the planet an "angry beast." You could also go with "war machine." Each day we arm it more.

THE ASSAULTS WILL NOT BE DISCRETE—THIS IS ANOTHER CLI-mate delusion. Instead, they will produce a new kind of cascading violence, waterfalls and avalanches of devastation, the planet pummeled again and again, with increasing intensity and in ways that build on each other and undermine our ability to respond, uprooting much of the landscape we have taken for granted, for centuries, as the stable foundation on which we walk, build homes and highways, shepherd our children through schools and into adulthood under the promise of safety—and subverting the promise that the world we have engineered and built for ourselves, out of nature, will also protect us against it, rather than conspiring with disaster against its makers.

Consider those California wildfires. In March 2018, Santa Barbara County issued mandatory evacuation orders for those living in Montecito, Goleta, Santa Barbara, Summerland, and Carpinteria—where the previous December's fires had hit hardest. It was the fourth evacuation order precipitated by a climate event in the county in just three months, but only the first had been for fire. The others were for mudslides ushered into possibility by that fire, one of the toniest communities in the most glamorous state of the world's preeminently powerful country upended by fear that their toy vineyards and hobby stables, their world-class beaches and lavishly funded public schools, would be inundated by rivers of mud, the community as thoroughly ravaged as the sprawling camps of temporary shacks housing Rohingya refugees from Myanmar in the monsoon region of Bangladesh. It was. More than a dozen died, including a toddler swept away by mud and carried miles down the mountainslope to the sea; schools closed and highways flooded, foreclosing the routes of emergency vehicles and making the community an inland island, as if behind a blockade, choked off by a mud noose.

Some climate cascades will unfold at the global level—cascades so large their effects will seem, by the curious legerdemain of environmental change, imperceptible. A warming planet leads to melting Arctic ice, which means less sunlight reflected back to the sun and more absorbed by a planet warming faster still, which means an ocean less able to absorb atmospheric carbon and so a planet warming faster still. A warming planet will also melt Arctic permafrost, which contains 1.8 trillion tons of carbon, more than twice as much as is currently suspended in the earth's atmosphere, and some of which, when it thaws and is released, may evaporate as methane, which is thirty-four times as powerful a greenhouse-gas warming blanket as carbon dioxide when judged on the timescale of a century; when judged on the timescale of two decades, it is eighty-six times as powerful. A hotter planet is, on net, bad for plant life, which means what is called "forest dieback"—the decline and retreat of jungle basins as big as countries and woods that sprawl for so many miles they used to contain whole folklores—which means a dramatic stripping-back of the planet's natural ability to absorb carbon and turn it into oxygen, which means still hotter temperatures, which means more dieback, and so on. Higher temperatures means more forest fires means fewer trees means less carbon absorption, means more carbon in the atmosphere, means a hotter planet still—and so on. A warmer planet means more water vapor in the atmosphere, and, water vapor being a greenhouse gas, this brings higher temperatures still—and so on. Warmer oceans can absorb less heat, which means more stays in the air, and contain less oxygen, which is doom for phytoplankton— which does for the ocean what plants do on land, eating carbon and producing oxygen—which leaves us with more carbon, which heats the planet further. And so on. These are the systems climate scientists call "feedbacks"; there are more. Some work in the other direction, moderating climate change. But many more point toward an acceleration of warming, should we trigger them. And just how these complicated, countervailing systems will interact—what effects will be exaggerated and what undermined by feedbacks—is unknown, which pulls a dark cloud of uncertainty over any effort to plan ahead for the climate future. We know what a best-case outcome for climate change looks like, however unrealistic, because it quite closely resembles the world as we live on it today. But we have not yet begun to contemplate those cascades that may bring us to the infernal range of the bell curve.

Other cascades are regional, collapsing on human communities and buckling them where they fall. These can be literal cascades—human-triggered avalanches are on the rise, with 50,000 people killed by avalanches globally between 2004 and 2016. In Switzerland, climate change has unleashed a whole new kind, thanks to what are called "rain-on-snow" events, which also caused the overflow of the Oroville Dam in Northern California and the 2013 flood of Alberta, Canada, with damages approaching $5 billion. But there are other kinds of cascade, too. Climate-driven water shortages or crop failures push climate refugees into nearby regions already struggling with resource scarcity. Sea-level rise inundates cropland with more and more saltwater flooding, transforming agricultural areas into brackish sponges no longer able to adequately feed those living off them; flooding power plants, knocking regions offline just as electricity may be needed most; and crippling chemical and nuclear plants, which, malfunctioning, breathe out their toxic plumes. The rains that followed the Camp Fire flooded the tent cities hastily assembled for the first disaster's refugees. In the case of the Santa Barbara mudslides, drought produced a state full of dry brush ripe for a spark; then a year of anomalously monsoonish rain produced only more growth, and wildfires tore through the landscape, leaving a mountainside without much plant life to hold in place the millions of tons of loose earth that make up the towering coastal range where the clouds tend to gather and the rain first falls.

Some of those watching from afar wondered, incredulously, how a mudslide could kill so many. The answer is, the same way as hurricanes or tornadoes—by weaponizing the environment, whether "man-made" or "natural." Wind disasters do not kill by wind, however brutal it gets, but by tugging trees out of earth and transforming them into clubs, making power lines into loose whips and electrified nooses, collapsing homes on cowering residents, and turning cars into tumbling boulders. And they kill slowly, too, by cutting off food delivery and medical supplies, making roads impassable even to first responders, knocking out phone lines and cell towers so that the ill and elderly must suffer, and hope to endure, in silence and without aid.

Most of the world is not Santa Barbara, with its Mission-style impasto of infinite-seeming wealth, and in the coming decades many of the most punishing climate horrors will indeed hit those least able to respond and recover. This is what is often called the problem of environmental justice; a sharper, less gauzy phrase would be "climate caste system." The problem is acute within countries, even wealthy ones, where the poorest are those who live in the marshes, the swamps, the floodplains, the inadequately irrigated places with the most vulnerable infrastructure—altogether an unwitting environmental apartheid. Just in Texas, 500,000 poor Latinos live in shantytowns called "colonias" with no drainage systems to deal with increased flooding.

The cleavage is even sharper globally, where the poorest countries will suffer more in our hot new world. In fact, with one exception—Australia—countries with lower GDPs will warm the most. That is notwithstanding the fact that much of the global south has not, to this point, defiled the atmosphere of the planet all that much. This is one of the many historical ironies of climate change that would better be called cruelties, so merciless is the suffering they will inflict. But disproportionately as it will fall on the world's least, the devastation of global warming cannot be easily quarantined in the developing world, as much as those in the Northern Hemisphere would probably, and not to our credit, prefer it. Climate disaster is too indiscriminate for that.

In fact, the belief that climate could be plausibly governed, or managed, by any institution or human instrument presently at hand is another wide-eyed climate delusion. The planet survived many millennia without anything approaching a world government, in fact endured nearly the entire span of human civilization that way, organized into competitive tribes and fiefdoms and kingdoms and nation-states, and only began to build something resembling a co-operative blueprint, very piecemeal, after brutal world wars—in the form of the League of Nations and United Nations and European Union and even the market fabric of globalization, whatever its flaws still a vision of cross-national participation, imbued with the neoliberal ethos that life on Earth was a positive-sum game. If you had to invent a threat grand enough, and global enough, to plausibly conjure into being a system of true international coop-eration, climate change would be it—the threat everywhere, and overwhelming, and total. And yet now, just as the need for that kind of cooperation is paramount, indeed necessary for anything like the world we know to survive, we are only unbuilding those alliances—recoiling into nationalistic corners and retreating from collective responsibility and from each other. That collapse of trust is a cascade, too.

JUST HOW COMPLETELY THE WORLD BELOW OUR FEET WILL BE-come unknown to us is not yet clear, and how we register its transfor-mation remains an open question. One legacy of the environmentalist creed that long prized the natural world as an otherworldly retreat is that we see its degradation as a sequestered story, unfolding sepa-rately from our own modern lives—so separately that the degrada-tion acquires the comfortable contours of parable, like pages from Aesop, aestheticized even when we know the losses as tragedy.

Climate change could soon mean that, in the fall, trees may simply turn brown, and so we will look differently at entire schools of painting, which stretched for generations, devoted to best cap-turing the oranges and reds we can no longer see ourselves out the windows of our cars as we drive along our highways. The coffee plants of Latin America will no longer produce fruit; beach homes will be built on higher and higher stilts and still be drowned. In many cases, it is better to use the present tense. In just the last forty years, according to the World Wildlife Fund, more than half of the world's vertebrate animals have died; in just the last twenty-five, one study of German nature preserves found, the flying insect population declined by three-quarters. The delicate dance of flowers and their pollinators has been disrupted, as have the migration patterns of cod, which have fled up the Eastern Seaboard toward the Arctic, evading the communities of fishermen that fed on them for centuries; as have the hibernation patterns of black bears, many of which now stay awake all winter. Species individuated over millions of years of evolution but forced together by climate change have begun to mate with one another for the first time, producing a whole new class of hybrid species: the pizzly bear, the coy-wolf. The zoos are already natural history museums, the children's books already out of date.

Older fables, too, will be remade: the story of Atlantis, having endured and enchanted for several millennia, will compete with the real-time sagas of the Marshall Islands and Miami Beach, each sinking over time into snorkelers' paradises; the strange fantasy of Santa and his polar workshop will grow eerier still in an Arctic of ice-free summers; and there is a terrible poignancy in contemplating how desertification of the entire Mediterranean Basin will change our reading of the *Odyssey*, or how it will discolor the shine of Greek islands for dust from the Sahara to permanently blanket their skies, or how it will recast the meaning of the Pyramids for the Nile to be dramatically drained. We will think of the border with Mexico differently, presumably, when the Rio Grande is a line traced through a dry riverbed—the Rio Sand, it's already been called. The imperious West has spent five centuries looking down its nose at the plight of those living within the pale of tropical disease, and one wonders how that will change when mosquitoes carrying malaria and dengue are flying through the streets of Copenhagen and Chicago, too.

But we have for so long understood stories about nature as allegories that we seem unable to recognize that the meaning of climate change is not sequestered in parable. It encompasses us; in a very real way it governs us—our crop yields, our pandemics, our migration patterns and civil wars, crime waves and domestic assaults, hurricanes and heat waves and rain bombs and megadroughts, the shape of our economic growth and everything that flows downstream from it, which today means nearly everything. Eight hundred million in South Asia alone, the World Bank says, would see their living conditions sharply diminish by 2050 on the current emissions track, and perhaps a climate slowdown will even reveal the bounty of what Andreas Malm calls fossil capitalism to be an illusion, sustained over just a few centuries by the arithmetic of adding the energy value of burned fossil fuels to what had been, before wood and coal and oil, an eternal Malthusian trap. In which case, we would have to retire the intuition that history will inevitably extract material progress from the planet, at least in any reliable or global pattern, and come to terms, somehow, with just how pervasively that intuition ruled even our inner lives, often tyrannically.

Adaptation to climate change is often viewed in terms of market trade-offs, but in the coming decades the trade will work in the opposite direction, with relative prosperity a benefit of more aggressive action. Every degree of warming, it's been estimated, costs a temperate country like the United States about one percentage point of GDP, and according to one recent paper, at 1.5 degrees the world would be $20 trillion richer than at 2 degrees. Turn the dial up another degree or two, and the costs balloon—the compound interest of environmental catastrophe. 3.7 degrees of warming would produce $551 trillion in damages, research suggests; total worldwide wealth is today about $280 trillion. Our current emissions trajectory takes us over 4 degrees by 2100; multiply that by that 1 percent of GDP and you have almost entirely wiped out the very possibility of economic growth, which has not topped 5 percent globally in over forty years. A fringe group of alarmed academics call this prospect "steady-state economics," but it ultimately suggests a more complete retreat from economics as an orienting beacon, and from growth as the lingua franca through which modern life launders all of its aspirations. "Steady-state" also gives a name to the creeping panic that history may be less progressive, as we've come to believe really only over the last several centuries, than cyclical, as we were sure it was for the many millennia before. More than that: in the vision steady-state economics projects of a state-of-nature competitive scramble, everything from politics to trade and war seems brutally zero-sum.

FOR CENTURIES WE HAVE LOOKED TO NATURE AS A MIRROR onto which to first project, then observe, ourselves. But what is the moral? There is nothing to learn from global warming, because we do not have the time, or the distance, to contemplate its lessons; we are after all not merely telling the story but living it. That is, trying to; the threat is immense. How immense? One 2018 paper sketches the math in horrifying detail. In the journal *Nature Climate Change,* a team led by Drew Shindell tried to quantify the suffering that would be avoided if warming was kept to 1.5 degrees, rather than 2 degrees—in other words, how much additional suffering would result from just that additional half-degree of warming. Their answer: 150 million more people would die from air pollution alone in a 2-degree warmer world than in a 1.5-degree warmer one. Later that year, the IPCC raised the stakes further: in the gap between 1.5 degrees and 2, it said, hundreds of millions of lives were at stake.

Numbers that large can be hard to grasp, but 150 million is the equivalent of twenty-five Holocausts. It is three times the size of the death toll of the Great Leap Forward—the largest nonmilitary death toll humanity has ever produced. It is more than twice the greatest death toll of any kind, World War II. The numbers don't begin to climb only when we hit 1.5 degrees, of course. As should not surprise you, they are already accumulating, at a rate of at least seven million deaths, from air pollution alone, each year—an annual Holocaust, pursued and prosecuted by what brand of nihilism?

This is what is meant when climate change is called an "existential crisis"—a drama we are now haphazardly improvising between

two hellish poles, in which our best-case outcome is death and suffering at the scale of twenty-five Holocausts, and the worst-case
outcome puts us on the brink of extinction. Rhetoric often fails
us on climate because the only factually appropriate language is of
a kind we've been trained, by a buoyant culture of sunny-side-up
optimism, to dismiss, categorically, as hyperbole.

Here, the facts are hysterical, and the dimensions of the drama
that will play out between those poles incomprehensibly large—
large enough to enclose not just all of present-day humanity but
all of our possible futures, as well. Global warming has improb-
ably compressed into two generations the entire story of human
civilization. First, the project of remaking the planet so that it is
undeniably ours, a project whose exhaust, the poison of emissions,
now casually works its way through millennia of ice so quickly you
can see the melt with a naked eye, destroying the environmental
conditions that have held stable and steadily governed for literally
all of human history. That has been the work of a single generation.
The second generation faces a very different task: the project of preserving our collective future, forestalling that devastation and engineering an alternate path. There is simply no analogy to draw on,
outside of mythology and theology—and perhaps the Cold War
prospect of mutually assured destruction.

Few feel like gods in the face of warming, but that the totality
of climate change should make us feel so passive—that is another
of its delusions. In folklore and comic books and church pews and
movie theaters, stories about the fate of the earth often perversely
counsel passivity in their audiences, and perhaps it should not surprise us that the threat of climate change is no different. By the
end of the Cold War, the prospect of nuclear winter had clouded
every corner of our pop culture and psychology, a pervasive nightmare that the human experiment might be brought to an end by
two jousting sets of proud, rivalrous tacticians, just a few sets of
twitchy hands hovering over the planet's self-destruct buttons. The
threat of climate change is more dramatic still, and ultimately more

democratic, with responsibility shared by each of us even as we shiver in fear of it; and yet we have processed that threat only in parts, typically not concretely or explicitly, displacing certain anxieties and inventing others, choosing to ignore the bleakest features of our possible future and letting our political fatalism and technological faith blur, as though we'd gone cross-eyed, into a remarkably familiar consumer fantasy: that someone else will fix the problem for us, at no cost. Those more panicked are often hardly less complacent, living instead through climate fatalism as though it were climate optimism.

Over the last few years, as the planet's own environmental rhythms have seemed to grow more fatalistic, skeptics have found themselves arguing not that climate change isn't happening, since extreme weather has made that undeniable, but that its causes are unclear—suggesting that the changes we are seeing are the result of natural cycles rather than human activities and interventions. It is a very strange argument; if the planet is warming at a terrifying pace and on a horrifying scale, it should transparently concern us more, rather than less, that the warming is beyond our control, possibly even our comprehension.

That we know global warming is our doing should be a comfort, not a cause for despair, however incomprehensibly large and complicated we find the processes that have brought it into being; that we know we are, ourselves, responsible for all of its punishing effects should be empowering, and not just perversely. Global warming is, after all, a human invention. And the flip side of our real-time guilt is that we remain in command. No matter how out-of-control the climate system seems—with its roiling typhoons, unprecedented famines and heat waves, refugee crises and climate conflicts—we are all its authors. And still writing.

Some, like our oil companies and their political patrons, are more prolific authors than others. But the burden of responsibility is too great to be shouldered by a few, however comforting it is to think all that is needed is for a few villains to fall. Each of us imposes some suffering on our future selves every time we flip on a light switch, buy a plane ticket, or fail to vote. Now we all share the responsibility to write the next act. We found a way to engineer devastation, and we can find a way to engineer our way out of it— or, rather, engineer our way toward a degraded muddle, but one that nevertheless extends forward the promise of new generations finding their own way forward, perhaps toward some brighter environmental future.

Since I first began writing about warming, I've often been asked whether I see any reason for optimism. The thing is, I *am* optimistic. Given the prospect that humans could engineer a climate that is 6 or even 8 degrees warmer over the course of the next several centuries—large swaths of the planet unlivable by any definition we use today—that degraded muddle counts, for me, as an encouraging future. Warming of 3 or 3.5 degrees would unleash suffering beyond anything that humans have ever experienced through many millennia of strain and strife and all-out war. But it is not a fatalistic scenario; in fact, it's a whole lot better than where we are headed. And in the form of carbon capture technology, which would extract CO_2 from the air, or geoengineering, which would cool the planet by suspending gas in the atmosphere, or other now-unfathomable innovations, we may conjure new solutions, which could bring the planet closer to a state we would today regard as merely grim, rather than apocalyptic.

I've also often been asked whether it's moral to reproduce in this climate, whether it's responsible to have children, whether it is fair to the planet or, perhaps more important, to the children. As it happens, in the course of writing this book, I did have a child, Rocca. Part of that choice was delusion, that same willful blindness: I know there are climate horrors to come, some of which will inevitably be visited on my children—that is what it means for warming to be an all-encompassing, all-touching threat. But those horrors are not yet scripted. We are staging them by inaction, and by action can stop them. Climate change means some bleak prospects for the decades ahead, but I don't believe the appropriate response to that challenge is withdrawal, is surrender. I think you have to do everything you can to make the world accommodate dignified and flourishing life, rather than giving up early, before the fight has been lost or won, and acclimating yourself to a dreary future brought into being by others less concerned about climate pain. The fight is, definitively, not yet lost—in fact will never be lost, so long as we avoid extinction, because however warm the planet gets, it will always be the case that the decade that follows could contain more suffering or less. And I have to admit, I am also excited, for everything that Rocca and her sisters and brothers will see, will witness, will do. She will hit her child-rearing years around 2050, when we could have climate refugees in the many tens of millions; she will be entering old age at the close of the century, the end-stage bookmark on all of our projections for warming. In between, she will watch the world doing battle with a genuinely existential threat, and the people of her generation making a future for themselves, and the generations they bring into being, on this planet. And she won't just be watching it, she will be living it—quite literally the greatest story ever told. It may well bring a happy ending.

What cause is there for hope? Carbon hangs in the air for decades, with some of the most terrifying feedbacks unspooling over even longer time horizons—which gives warming the eerie shimmer of an unending menace. But climate change is not an ancient crime we are tasked with solving now; we are destroying our planet every day, often with one hand as we conspire to restore it with the other. Which means, as Paul Hawken has perhaps illustrated most coolheadedly, we can also stop destroying it, in the same style—collectively, haphazardly, in all the most quotidian ways in addition to the spectacular-seeming ones. The project of unplugging the entire industrial world from fossil fuels is intimidating, and must be done in fairly short order—by 2040, many scientists say. But in the meantime many avenues are open—wide open, if we are not too lazy and too blinkered and too selfish to embark upon them.

Fully half of British emissions, it was recently calculated, come from inefficiencies in construction, discarded and unused food, electronics, and clothing; two-thirds of American energy is wasted; globally, according to one paper, we are subsidizing the fossil fuel business to the tune of $5 trillion each year. None of that has to continue. Slow-walking action on climate, another paper found, will cost the world $26 trillion by just 2030. That does not have to continue. Americans waste a quarter of their food, which means that the carbon footprint of the average meal is a fourth larger than it has to be. That need not continue. Five years ago, hardly anyone outside the darkest corners of the internet had even heard of Bitcoin; today mining it consumes more electricity than is generated by all the world's solar panels combined, which means that in just a few years we've assembled, out of distrust of one another and the nations behind "fiat currencies," a program to wipe out the gains of several long, hard generations of green energy innovation. It did not have to be that way. And a simple change to the algorithm could eliminate that Bitcoin footprint entirely.

These are just a few of the reasons to believe that what the Canadian activist Stuart Parker has called "climate nihilism" is, in fact, another of our delusions. What happens, from here, will be entirely our own doing. The planet's future will be determined in large part by the arc of growth in the developing world—that's where most of the people are, in China and India and, increasingly, sub-Saharan Africa. But this is no absolution for the West, where the average citizen produces many times more emissions than almost anyone in Asia, just out of habit. I toss out tons of wasted food and hardly ever recycle; I leave my air-conditioning on; I bought into Bitcoin at the peak of the market. None of that is necessary, either.

But it also isn't necessary for Westerners to adopt the lifestyle of the global poor. Seventy percent of the energy produced by the planet, it's estimated, is lost as waste heat. If the average American were confined by the carbon footprint of her European counterpart, U.S. carbon emissions would fall by more than half. If the world's richest 10 percent were limited to that same footprint, global emissions would fall by a third. And why shouldn't they be? Almost as a prophylactic against climate guilt, as the news from science has grown bleaker, Western liberals have comforted themselves by contorting their own consumption patterns into performances of moral or environmental purity—less beef, more Teslas, fewer transatlantic flights. But the climate calculus is such that individual lifestyle choices do not add up to much, unless they are scaled by politics. America's rump climate party aside, that scaling should not be impossible, once we understand the stakes. In fact, the stakes mean, it must not be.

ANNIHILATION IS ONLY THE VERY THIN TAIL OF WARMING'S very long bell curve, and there is nothing stopping us from steering clear of it. But what lies between us and extinction is horrifying enough, and we have not yet begun to contemplate what it means to live under those conditions—what it will do to our politics and our culture and our emotional equilibria, our sense of history and our relationship to it, our sense of nature and our relationship to it, that we are living in a world degraded by our own hands, with the horizon of human possibility dramatically dimmed. We may yet see a climate deus ex machina—or, rather, we may yet build one, in the form of carbon capture technology or geoengineering, or in the form of a revolution in the way we generate power, electric or political. But that solution, if it comes at all, will emerge against a bleak horizon, darkened by our emissions as if by glaucoma.

Especially those who have imbibed several centuries of Western triumphalism tend to see the story of human civilization as an inevitable conquest of the earth, rather than the saga of an insecure culture, like mold, growing haphazardly and unsurely upon it. That fragility, which pervades now everything humans might do on this planet, is the great existential insight of global warming, but it is only beginning to shake our triumphalism—though, if we had stopped to contemplate the possibilities a generation ago, it probably would not surprise us to see a new form of political nihilism emerging in the region of the world already baked hardest by global warming, the Middle East, and expressed there through suicidal spasms of theological violence. That region was once called, grandly, "the cradle of civilization." Today, political nihilism radiates almost everywhere, through the many cultures that arose, branching, from Middle Eastern roots. We have all already left behind the narrow window of environmental conditions that allowed the human animal to evolve in the first place, but not just evolve—that window has enclosed everything we remember as history, and value as progress, and study as politics. What will it mean to live outside that window, probably quite far outside it? That reckoning is the subject of this book.

NONE OF IT IS NEWS. THE SCIENCE THAT MAKES UP THE FOL-lowing twelve chapters has been culled from interviews with dozens of experts, and from hundreds of papers published in the best academic journals over the previous decade or so. Since it is science, it is tentative, ever-evolving, and some of the predictions that follow will surely not come precisely to pass. But it is an honest and fair portrait of the state of our collective understanding of the many multiplying threats that a warming planet poses to all of us presently living on it, and hoping we may continue to do so, in an indefinite and undisturbed way.

Little of it is about "nature" per se, and none concerns the tragic fate of the planet's animals, which has been written about so elegantly and poetically by others that, like our sea-level myopia, it threatens to occlude our picture of what global warming means for us, the human animal. Until now, it seems to have been easier for us to empathize with the climate plight of other species than our own, perhaps because we have such a hard time acknowledging or understanding our own responsibility and complicity in the changes now unfolding, and such an easier time evaluating the morally simpler calculus of pure victimhood.

What follows is instead a kaleidoscopic accounting of the

human costs of human life continuing as it has for a generation, which will fill up the planet with only more humans—what ongoing global warming spells for public health, for conflict, for politics and food production and pop culture, for urban life and mental health and the way we imagine our own futures as we begin to perceive, all around us, an acceleration of history and the diminishing of possibility that acceleration likely brings. The force of retribution will cascade down to us through nature, but the cost to nature is only one part of the story; we will all be hurting. I may be in the minority in feeling that the world could lose much of what we think of as "nature," as far as I cared, so long as we could go on living as we have in the world left behind. The problem is, we can't.

II

Elements of Chaos

Heat Death

HUMANS, LIKE ALL MAMMALS, ARE HEAT ENGINES; SURVIV-ing means having to continually cool off, as panting dogs do. For that, the temperature needs to be low enough for the air to act as a kind of refrigerant, drawing heat off the skin so the engine can keep pumping. At seven degrees of warming, that would become impossible for portions of the planet's equatorial band, and especially the tropics, where humidity adds to the problem. And the effect would be fast: after a few hours, a human body would be cooked to death from both inside and out.

At eleven or twelve degrees Celsius of warming, more than half the world's population, as distributed today, would die of direct heat. Things almost certainly won't get that hot anytime soon, though some models of unabated emissions do bring us that far eventually, over centuries. But at just five degrees, according to some calculations, whole parts of the globe would be literally unsurvivable for humans. At six, summer labor of any kind would become impossible in the lower Mississippi Valley, and everybody in the United States east of the Rockies would suffer more from heat than anyone, anywhere, in the world today. New York City would be hotter than present-day Bahrain, one of the planet's hottest spots, and the temperature in Bahrain "would induce hyperthermia in even sleeping humans."

Five or six degrees is unlikely by 2100. The IPCC furnishes us with a median prediction of over four degrees, should we continue down the current emissions path. That would deliver what today seems like unthinkable impacts—wildfires burning sixteen times as much land in the American West, hundreds of drowned cities. Cities now home to millions, across India and the Middle East, would become so hot that stepping outside in summer would be a lethal risk—in fact, they will become that way much sooner, with as little as two degrees of warming. You do not need to consider worst-case scenarios to become alarmed.

With direct heat, the key factor is something called "wet-bulb temperature," which also measures humidity in a combined method as home-laboratory-kit as it sounds: the temperature is registered on a thermometer wrapped in a damp sock as it's swung around in the air. At present, most regions reach a wet-bulb maximum of 26 or 27 degrees Celsius; the true red line for habitability is 35 degrees, beyond which humans begin simply dying from the heat. That leaves a gap of 8 degrees. What is called "heat stress" comes much sooner.

Actually, we're there already. Since 1980, the planet has experienced a fiftyfold increase in the number of dangerous heat waves; a bigger increase is to come. The five warmest summers in Europe since 1500 have all occurred since 2002, and eventually, the IPCC warns, simply working outdoors at that time of year will be unhealthy for parts of the globe. Even if we meet the Paris goals, cities like Karachi and Kolkata will annually encounter deadly heat waves like those that crippled them in 2015, when heat killed thousands in India and Pakistan. At four degrees, the deadly European heat wave of 2003, which killed as many as 2,000 people a day, will be a normal summer. Then, it was one of the worst weather events in Continental history, killing 35,000 Europeans, including 14,000 French; perversely, the infirm fared relatively well, William Langewiesche has written, most of them watched over in the nursing homes and hospitals of those well-off countries, and it was the comparatively healthy elderly who accounted for most of the dead, many left behind by vacationing families escaping the heat, with some corpses rotting for weeks before the families returned.

It will get worse. In that "business as usual" scenario, a research team led by Ethan Coffel calculated in 2017, the number of days warmer than what were once the warmest days of the year could grow by a factor of 100 by 2080. Possibly by a factor of 250. The metric Coffel uses is "person-days": a unit that combines the number of people affected with the number of days. Every year, there would be between 150 and 750 million person-days with wet-bulb temperatures equivalent to today's most severe—i.e., quite deadly—heat waves. There would be a million person-days each year with intolerable wet-bulb temperatures—combinations of heat and humidity beyond the human capacity for survival. By the end of the century, the World Bank has estimated, the coolest months in tropical South America, Africa, and the Pacific are likely to be warmer than the warmest months at the end of the twentieth century.

We had heat waves back then, of course, deadly ones; in 1998, the Indian summer killed 2,500. More recently, temperature spikes have gotten hotter. In 2010, 55,000 died in a Russian heat wave that killed 700 people in Moscow each day. In 2016, in the midst of a heat wave that baked the Middle East for several months, temperatures in Iraq broke 100 degrees Fahrenheit in May, 110 in June, and 120 in July, with temperatures dipping below 100, most days, only at night. (A Shiite cleric in Najaf proclaimed the heat was the result of an electromagnetic attack on the country by American forces, according to *The Wall Street Journal*, and some state meteorologists agreed.) In 2018, the hottest temperature likely ever recorded in April was registered in southeast Pakistan. In India, a single day over 95 degrees Fahrenheit increases annual mortality rates by three-quarters of a percent; in 2016, a string of days topped 120—in May. In Saudi Arabia, where summer temperatures often approach that mark, 700,000 barrels of oil are burned each day in the summer, mostly to power the nation's air-conditioning.

That can help with the heat, of course, but air conditioners and fans already account for fully 10 percent of global electricity consumption. Demand is expected to triple, or perhaps quadruple, by 2050; according to one estimate, the world will be adding 700 million AC units by just 2030. Another study suggests that by 2050 there will be, around the world, more than nine billion cooling appliances of various kinds. But, the climate-controlled malls of the Arab emirates aside, it is not remotely economical, let alone "green," to wholesale air-condition all the hottest parts of the planet, many of them also the poorest. And indeed, the crisis will be most dramatic across the Middle East and Persian Gulf, where in 2015 the heat index registered temperatures as high as 163 degrees Fahrenheit. As soon as several decades from now, the hajj will become physically impossible for many of the two million Muslims who currently make the pilgrimage each year.

It is not just the hajj, and it is not just Mecca. In the sugarcane region of El Salvador, as much as one-fifth of the population—including over a quarter of the men—has chronic kidney disease, the presumed result of dehydration from working the fields they were able to comfortably harvest as recently as two decades ago. With dialysis, which is expensive, those with kidney failure can expect to live five years; without it, life expectancy is measured in weeks. Of course, heat stress promises to assail us in places other than our kidneys, too. As I type that sentence, in the California desert in mid-June, it is 121 degrees outside my door. It is not a record high.

THIS IS AMONG THE THINGS COSMOLOGISTS MEAN WHEN THEY talk about the utter improbability of anything as advanced as human intelligence evolving anywhere in a universe as inhospitable to life as this one: every uninhabitable planet out there is a reminder of just how unique a set of circumstances is required to produce a climate equilibrium supportive of life. No intelligent life that we know of ever evolved, anywhere in the universe, outside of the narrow Goldilocks range of temperatures that enclosed all of human evolution, and that we have now left behind, probably permanently.

How much hotter will it get? The question may sound scientific, inviting expertise, but the answer is almost entirely human—which is to say, political. The menace of climate change is a mercurial one; uncertainty makes it a shape-shifting threat. When will the planet warm by two degrees, and when by three? How much sea-level rise will be here by 2030, by 2050, by 2100, as our children are leaving the earth to their children and grandchildren? Which cities will flood, which forests will dry out, which breadbaskets will become husks? That uncertainty is among the most momentous metanarratives that climate change will bring to our culture over the next decades—an eerie lack of clarity about what the world we live in will even look like, just a decade or two down the road, when we will still be living in the same homes and paying the same mortgages, watching the same television shows and making appeals to many of the same justices of the Supreme Court. But while there are a few things science does not know about how the climate system will respond to all the carbon we've pumped into the air, the uncertainty of what will happen—that haunting uncertainty—emerges not from scientific ignorance but, overwhelmingly, from the open question of how we respond. That is, principally, how much more carbon we decide to emit, which is not a question for the natural sciences but the human ones. Climatologists can, today, predict with uncanny accuracy where a hurricane will hit, and at what intensity, as much as a week out from landfall; this is not just because the models are good but because all the inputs are known. When it comes to global warming, the models are just as good, but the key input is a mystery: What will we do?

The lessons there are unfortunately bleak. Three-quarters of a century since global warming was first recognized as a problem, we have made no meaningful adjustment to our production or consumption of energy to account for it and protect ourselves. For far too long, casual climate observers have watched scientists draw pathways to a stable climate and concluded that the world would adapt accordingly; instead, the world has done more or less nothing, as though those pathways would implement themselves. Market forces have delivered cheaper and more widely available green energy, but the same market forces have absorbed those innovations, which is to say profited from them, while continuing to grow emissions. Politics has produced gestures of tremendous global solidarity and cooperation, then discarded those promises immediately. It has become commonplace among climate activists to say that we have, today, all the tools we need to avoid catastrophic climate change—even major climate change. It is also true. But political will is not some trivial ingredient, always at hand. We have the tools we need to solve global poverty, epidemic disease, and abuse of women, as well.

It was as recent as 2016 that the celebrated Paris climate accords were adopted—defining two degrees of global warming as a must-meet target and rallying all the world's nations to meet it—and the returns are already dispiritingly grim. In 2017, carbon emissions grew by 1.4 percent, according to the International Energy Agency, after an ambiguous couple of years optimists had hoped represented a leveling-off, or peak; instead, we're climbing again. Even before the new spike, not a single major industrial nation was on track to fulfill the commitments it made in the Paris treaty. Of course, those commitments only get us down to 3.2 degrees; to keep the planet under 2 degrees of warming, all signatory nations have to significantly better their pledges. At present, there are 195 signatories, of which only the following are considered even "in range" of their Paris targets: Morocco, Gambia, Bhutan, Costa Rica, Ethiopia, India, and the Philippines. This puts Donald Trump's commitment to withdraw from the treaty in a useful perspective; in fact, his spite may ultimately prove perversely productive, since the evacuation of American leadership on climate seems to have mobilized China—giving Xi Jinping an opportunity and an enticement to adopt a much more aggressive posture toward climate. Of course those renewed Chinese commitments are, at this point, just rhetorical, too; the country already has the world's largest footprint, and in the first three months of 2018 its emissions grew by 4 percent. China commands half of the planet's coal-power capacity, with plants that only operate, on average, half of the time—which means their use could quickly grow. Globally, coal power has nearly doubled since 2000. According to one analysis, if the world as a whole followed the Chinese example, it would bring five degrees of warming by 2100.

In 2018, the United Nations predicted that at the current emissions rate the world would pass 1.5 degrees by 2040, if not sooner; according to the 2017 National Climate Assessment, even if global carbon concentration was immediately stabilized, we should expect more than half a degree Celsius of additional warming to come. Which is why staying below 2 degrees probably requires not just carbon scale-back but what are called "negative emissions." These tools come in two forms: technologies that would suck carbon out of the air (called CCS, for "carbon capture and storage") and new approaches to forestry and agriculture that would do the same, in a slightly more old-fashioned way (bioenergy with carbon capture and storage, or "BECCS").

According to a raft of recent papers, both are something close to fantasy, at least at present. In 2018, the European Academies' Science Advisory Council found that existing negative-emissions technologies have "limited realistic potential" to even slow the increase in concentration of carbon in the atmosphere—let alone meaningfully reduce that concentration. In 2018, *Nature* dismissed all scenarios built on CCS as "magical thinking." It is not even so pleasant to engage in that thinking. There is not much carbon in the air, all told, just 410 parts per million, but it is everywhere, and so relying on carbon capture globally could require large-scale scrubbing plantations nearly everywhere on Earth—the planet transformed into something like an air-recycling plant orbiting the sun, an industrial satellite tracing a parabola through the solar system. (This is not what Barbara Ward or Buckminster Fuller meant by "spaceship earth.") And while advances are sure to come, bringing costs down and making more efficient machines, we can't wait much longer for that progress; we simply don't have the time. One estimate suggests that, to have hopes of two degrees, we need to open new full-scale carbon capture plants at the pace of one and a half per day every day for the next seventy years. In 2018, the world had eighteen of them, total.

This is not good, but indifference is unfortunately nothing new when it comes to climate. Projecting future warming is a foolish game, given how many layers of uncertainty govern the outcome; but if a best-case scenario is now somewhere between 2 and 2.5 degrees of warming by 2100, it seems that the likeliest outcome, the fattest part of the bell curve of probability, sits at about 3 degrees, or just a bit above. Probably even that amount of warming would require significant negative-emissions use, given that our use of carbon is still growing. And there is also some risk from scientific uncertainty, the possibility that we are underestimating the effects of those feedback loops in natural systems we only poorly understand. Conceivably, if those processes are triggered, we could hit 4 degrees of warming by 2100, even with a meaningful reduction in emissions over the coming decades. But the track record since Kyoto implies that human shortsightedness makes it unproductive to offer predictions about what *will* happen, when it comes to emissions and warming; better to consider what *could* happen. The sky is literally the limit.

CITIES, WHERE THE WORLD WILL OVERWHELMINGLY LIVE IN the near future, only magnify the problem of high temperature. Asphalt and concrete and everything else that makes a city dense, including human flesh, absorb ambient heat, essentially storing it for a time like a slow-release poison pill; this is especially problematic because, in a heat wave, nightly reprieves are vital, allowing bodies to recover. When those reprieves are shorter, and shallower, flesh simply continues to simmer. In fact, the concrete and asphalt of cities absorb so much heat during the day that when it is released, at night, it can raise the local temperature as much as 22 degrees Fahrenheit, turning what could be bearably hot days into deadly ones—as in the Chicago heat wave of 1995, which killed 739 people, the direct-heat effects compounded by broken public health infrastructure. That commonly cited figure only reflects immediate deaths; of the many thousands more who visited hospitals during the heat wave, almost half died within the year. Others merely suffered permanent brain damage. Scientists call this the "heat island" effect—each city its own enclosed space, and hotter the more crowded it is.

Of course, the world is rapidly urbanizing, with the United Nations estimating that two-thirds of the global population will live in cities by 2050—2.5 billion new urbanites, by that count. For a century or more, the city has seemed like a vision of the future to much of the world, which keeps inventing new scales of metropolis: bigger than 5 million people, bigger than 10, bigger than 20. Climate change won't likely slow that pattern by much, but it will make the great migrations it reflects more perilous, with many millions of the world's ambitious flooding into cities whose calendars are dotted with days of deadly heat, gathering in those new megalopolises like moths to a flame.

In theory, climate change could even reverse those migrations, perhaps more totally than crime did in many American cities in the last century, turning urban populations in certain parts of the world outward as the cities themselves become unbearable. In the heat, roads in cities will melt and train tracks will buckle—this is actually happening already, but the impacts will mushroom in the decades ahead. Currently, there are 354 major cities with average maximum summertime temperatures of 95 degrees Fahrenheit or higher. By 2050, that list could grow to 970, and the number of people living in those cities and exposed to that deadly heat could grow eightfold, to 1.6 billion. In the United States alone, 70,000 workers have been seriously injured by heat since 1992, and by 2050, 255,000 are expected to die globally from direct heat effects. Already, as many as 1 billion are at risk for heat stress worldwide, and a third of the world's population is subject to deadly heat waves at least twenty days each year; by 2100, that third will grow to half, even if we manage to pull up short of two degrees. If we don't, the number could climb to three-quarters.

In the United States, heat stroke has a pathetic reputation—a plague you learn about from summer camp, like swimming cramps. But heat death is among the cruelest punishments to a human body, just as painful and disorienting as hypothermia. First comes "heat exhaustion," mostly a mark of dehydration: profuse sweating, nausea, headache. After a certain point, though, water won't help, your core temperature rising as your body sends blood outward to the skin, hoping desperately to cool it down. The skin often reddens; internal organs begin to fail. Eventually you could stop sweating. The brain, too, stops working properly, and sometimes, after a period of agitation and combativeness, the episode is punctuated with a lethal heart attack. "When it comes to extreme heat," Langewiesche has written, "you can no more escape the conditions than you can shed your skin."

Hunger

CLIMATES DIFFER AND PLANTS VARY, BUT THE BASIC RULE of thumb for staple cereal crops grown at optimal temperature is that for every degree of warming, yields decline by 10 percent. Some estimates run higher. Which means that if the planet is five degrees warmer at the end of the century, when projections suggest we may have as many as 50 percent more people to feed, we may also have 50 percent less grain to give them. Or even less, because yields actually decline faster the warmer things get. And proteins are worse: it takes eight pounds of grain to produce just a single pound of hamburger meat, butchered from a cow that spent its life warming the planet with methane burps.

Globally, grain accounts for about 40 percent of the human diet; when you add soybeans and corn, you get up to two-thirds of all human calories. Overall, the United Nations estimates that the planet will need nearly twice as much food in 2050 as it does today—and although this is a speculative figure, it's not a bad one.

Pollyannaish plant physiologists will point out that the cereal-crop math applies only to those regions already at peak growing temperature, and they are right—theoretically, a warmer climate will make it easier to grow wheat in Greenland. But as a pathbreaking paper by Rosamond Naylor and David Battisti pointed out, the tropics are already too hot to efficiently grow grain, and those places where grain is produced today are already at optimal growing temperature—which means even a small warming will push them down a slope of declining productivity. The same, broadly speaking, is true for corn. At four degrees of warming, corn yields in the United States, the world's top producer of maize, are expected to drop by almost half. Predicted declines are not quite as dramatic in the next three biggest producers—China, Argentina, Brazil—but in each case the country would lose at least a fifth of its productivity.

A decade ago, climatologists might have told you that although direct heat undermined plant growth, the extra carbon in the atmosphere would have the opposite effect—a kind of airborne fertilizer. The effect is strongest on weeds, though, and does not seem to hold for grain. And at higher concentrations of carbon, plants grow thicker leaves, which sounds innocuous. But thicker leaves are worse at absorbing CO_2, an effect that means, by the end of the century, as much as 6.39 billion additional tons in the atmosphere each year.

Beyond carbon, climate change means staple crops are doing battle with more insects—their increased activity could cut yields an additional 2 to 4 percent—as well as fungus and disease, not to mention flooding. Some crops, like sorghum, are a bit more robust, but even in those regions where such alternatives have been a staple, their production has diminished recently; and while grain breeders have some hope that they can produce more heat-tolerant strains, they've been trying for decades without success. The world's natural wheat belt is moving poleward by about 160 miles each decade, but you can't easily move croplands north a few hundred miles, and not just because it's difficult to suddenly clear the land occupied now by towns, highways, office parks, and industrial installations. Yields in places like remote areas of Canada and Russia, even if they warmed by a few degrees, would be limited by the quality of soil there, since it takes many centuries for the planet to produce optimally fertile dirt. The lands that are fertile are the ones we are already using, and the climate is changing much too fast to wait for the northern soil to catch up. That soil, believe it or not, is literally disappearing—75 billion tons of soil lost each year. In the United States, the rate of erosion is ten times as high as the natural replenishment rate; in China and India, it is thirty to forty times as fast.

Even when we try to adapt, we move too slowly. Economist Richard Hornbeck specializes in the history of the American Dust Bowl; he says that farmers of that era could conceivably have adapted to the changing climate of their time by cultivating different crops. But they didn't, lacking credit to make the necessary investments—and were therefore unable to shake inertia and ritual and the rootedness of identity. So instead the crops died out, in cascading waves crashing through whole American states and all the people living in them.

As it happens, a similar transformation is unfolding in the American West right now. In 1879, the naturalist John Wesley Powell, who spent his downtime as a soldier during the Battle of Vicksburg studying the rocks that filled the Union trenches, divined a natural boundary running due north along the 100th meridian. It separated the humid—and therefore cultivatable—natural farmland of what became the Midwest from the arid, spectacular, but less farmable land of the true West. The divide ran through Texas, Oklahoma, Kansas, Nebraska, and the Dakotas, and stretches south into Mexico and north into Manitoba, Canada, separating more densely populated communities full of large farms from sparser, open land that was never truly made valuable by agriculture. Since just 1980, that boundary has moved fully 140 miles east, almost to the 98th parallel, drying up hundreds of thousands of square miles of farmland in the process. The planet's only other similar boundary is the one separating the Sahara desert from the rest of Africa. That desert has expanded by 10 percent, too; in the winter, the figure is 18 percent.

THE PRIVILEGED CHILDREN OF THE INDUSTRIALIZED WEST have long laughed at the predictions of Thomas Malthus, the British economist who believed that long-term economic growth was impossible, since each bumper crop or episode of growth would ultimately produce more children to consume or absorb it—and as a result the size of any population, including that of the planet as a whole, was a check against material well-being. In 1968, Paul Ehrlich made a similar warning, updated for a twenty-first-century planet with many times more people on it, with his widely derided *The Population Bomb*, which proposed that the economic and agricultural productivity of the earth had already reached its natural limit—and which was published, as it happened, just as the productivity gains from what's called the "green revolution" were coming into focus. That term, which today is sometimes used to describe advances in clean energy, first arose to name the incredible boom in agricultural yields produced by innovations in farming practices in the middle of the twentieth century. In the half century since, not only has the world's population doubled but the fraction of people living in extreme poverty has fallen by a factor of about six—from just more than half of humanity to 10 percent. In the world's developing countries, undernourishment has dropped from more than 30 percent in 1970 to close to 10 percent today.

These developments counsel sanguinity in the face of all kinds of environmental pressures, and in his recent book on the meaning of the twentieth-century agricultural boom, the writer Charles Mann divides those who respond to the seeming challenge of resource scarcity with reflexive optimism, whom he calls "wizards," from those who see collapse always around the corner, whom he calls "prophets." But though the green revolution seems almost too perfectly conceived and executed to refute Ehrlich's alarmism, Mann himself is not sure what the lessons are. It may yet be a bit early to judge Ehrlich—or perhaps even his godfather, Malthus—since nearly all of the astonishing productivity gains of the last century trace back to the work of a single man, Norman Borlaug, perhaps the best argument for the humanitarian virtue of America's imperial century. Born to Iowa family farmers in 1914, he went to state school, found work at DuPont, and then, with the help of the Rockefeller Foundation, developed a new collection of high-yield, disease-resistant wheat varieties that are now credited with saving the lives of a billion people worldwide. Of course, if those gains were a onetime boost—engineered, in large part, by a single man—how comfortably can we count on future improvements?

The academic term for the subject of this debate is "carrying capacity": How much population can a given environment ultimately support before collapsing or degrading from overuse? But it is one thing to consider what might be the maximum yield of a particular plot of earth and another to contemplate how fully that number is governed by environmental systems—systems far larger and more diffusely determined than even an imperial wizard like Borlaug could reasonably expect to command and control. Global warming, in other words, is more than just one input in an equation to determine carrying capacity; it is the set of conditions under which all of our experiments to improve that capacity will be conducted. In this way, climate change appears to be not merely one challenge among many facing a planet already struggling with civil strife and war and horrifying inequality and far too many other insoluble hardships to iterate, but the all-encompassing stage on which all those challenges will be met—a whole sphere, in other words, which literally contains within it all of the world's future problems and all of its possible solutions.

Curiously, maddeningly, these can be the same. The graphs that show so much recent progress in the developing world—on poverty, on hunger, on education and infant mortality and life expectancy and gender relations and more—are, practically speaking, the same graphs that trace the dramatic rise in global carbon emissions that has brought the planet to the brink of overall catastrophe. This is one aspect of what is meant by the term "climate justice." Not only is it undeniably the case that the cruelest impacts of climate change will be borne by those least resilient in the face of climate tragedy, but to a large degree what could be called the humanitarian growth of the developing world's middle class since the end of the Cold War has been paid for by fossil-fuel-driven industrialization—an investment in the well-being of the global south made by mortgaging the ecological future of the planet.

This is one reason that our global climate fate will be shaped so overwhelmingly by the development patterns of China and India, who have the tragic burden of trying to bring many hundreds of millions more into the global middle class while knowing that the easy paths taken by the nations that industrialized in the nineteenth and even twentieth centuries are now paths to climate chaos. Which is not to say they won't follow them anyway: by 2050, milk consumption in China is expected to grow to triple the current level, thanks to the changing, West-facing tastes of its emerging consumer classes, a single-item boom in a single country that is expected, all by itself, to increase global greenhouse-gas emissions from dairy farming by about 35 percent.

Already, global food production accounts for about a third of all emissions. To avoid dangerous climate change, Greenpeace has estimated that the world needs to cut its meat and dairy consumption in half by 2050; everything we know about what happens when countries get wealthier suggests this will be close to impossible. And turning away from milk is one thing; turning down cheap electrification, automobile culture, or the protein-heavy diets the world's wealthy rely on to stay thin are much bigger asks. In the postindustrial West, we try not to think about these bargains, which have benefited us so enormously. When we do, it is often in the guilty spirit of what critic Kris Bartkus has memorably called "the Malthusian tragic"—namely, our inability to see any remaining innocence in the quotidian life of the well-to-do West, given the devastation that wealth has imposed on the world of natural wonder it conquered and the suffering of those, elsewhere on the planet, left behind in the race to endless material comforts. And asked, functionally, to pay for them.

Of course, most have not embraced that tragic, or self-pitying, view. A state of half-ignorance and half-indifference is a much more pervasive climate sickness than true denial or true fatalism. It is the subject of William Vollmann's grand, two-part *Carbon Ideologies*, which opens—beyond the epigraph "A crime is something someone else commits," from Steinbeck—like this: "Someday, perhaps not long from now, the inhabitants of a hotter, more dangerous and biologically diminished planet than the one on which I lived may wonder what you and I were thinking, or whether we thought at all." For much of the book's prologue, he writes in a past tense rendered from an imagined, devastated future. "Of course we did it to ourselves; we had always been intellectually lazy, and the less asked of us, the less we had to say," he writes. "We all lived for money, and that is what we died for."

DROUGHT MAY BE AN EVEN BIGGER PROBLEM FOR FOOD PRO-duction than heat, with some of the world's most arable land turning quickly to desert. At 2 degrees of warming, droughts will wallop the Mediterranean and much of India, and corn and sorghum all around the world will suffer, straining global food supply. At 2.5 degrees, thanks mostly to drought, the world could enter a global food deficit—needing more calories than the planet can produce. At 3 degrees, there would be further drought—in Central America, Pakistan, the western United States, and Australia. At 5 degrees, the whole earth would be wrapped in what the environmentalist Mark Lynas calls "two globe-girdling belts of perennial drought."

Precipitation is notoriously hard to model in detail, yet predictions for later this century are basically unanimous: both unprecedented droughts and unprecedented flood-producing rains. By 2080, without dramatic reductions in emissions, southern Europe will be in permanent extreme drought, much worse than the American Dust Bowl ever was. The same will be true in Iraq and Syria and much of the rest of the Middle East; some of the most densely populated parts of Australia, Africa, and South America; and the breadbasket regions of China. None of these places, which today supply much of the world's food, would be reliable sources going forward. As for the original Dust Bowl: the droughts in the American plains and Southwest would not just be worse than in the 1930s, a 2015 NASA study predicted, but worse than any droughts in a thousand years—and that includes those that struck between 1100 and 1300, which dried up all the rivers east of the Sierra Nevada mountains and may have been responsible for the death of the Anasazi civilization.

Remember, even with the remarkable gains of the last decades, we do not presently live in a world without hunger. Far from it: most estimates put the number of undernourished at 800 million globally, with as many as 100 million hungry because of climate shocks. What is called "hidden hunger"—micronutrient and dietary deficiencies—is considerably higher, affecting well over 1 billion people. The spring of 2017 brought an unprecedented quadruple famine to Africa and the Middle East; the United Nations warned that those separate starvation events in Somalia, South Sudan, Nigeria, and Yemen could kill 20 million that year. That was a single year in a single region. Africa is today straining to feed about 1 billion people, a population expected to quadruple over the course of the twenty-first century to 4 billion.

One hopes these population booms will bring their own Borlaugs, ideally many of them. And already there are some hints of possible technological breakthroughs: China has invested in truly customized farming strategies to boost productivity and cut the use of greenhouse-gas-producing fertilizer; in Britain, a "soil-free startup" announced its first "harvest" in 2018; in the United States, you already hear about the prospects for vertical farming, which saves farmland by stacking crops indoors; and lab-grown protein, which does the same by culturing meats inside test tubes. But these remain vanguard technologies, distributed unequally and, being so expensive, unavailable for now to the many who are most in need. A decade ago, there was great optimism that GMO crops could produce another green revolution, but today gene modification has been used mostly to make plants more resistant to pesticides, pesticides manufactured and sold by the same companies engineering the crops. And cultural resistance has grown so rapidly that Whole Foods now advertises its house brand of seltzer as "GMO-free sparkling water."

It is far from clear how much benefit even those able to take advantage of vanguard techniques will be able to reap. Over the past fifteen years, the iconoclastic mathematician Irakli Loladze has isolated a dramatic effect of carbon dioxide on human nutrition unanticipated by plant physiologists: it can make plants bigger, but those bigger plants are less nutritious. "Every leaf and every grass blade on earth makes more and more sugars as CO_2 levels keep rising," Loladze told *Politico,* in a story about his work headlined "The Great Nutrient Collapse." "We are witnessing the greatest injection of carbohydrates into the biosphere in human history—[an] injection that dilutes other nutrients in our food supply."

Since 1950, much of the good stuff in the plants we grow—protein, calcium, iron, vitamin C, to name just four—has declined by as much as one-third, a landmark 2004 study showed. Everything is becoming more like junk food. Even the protein content of bee pollen has dropped by a third.

The problem has gotten worse as carbon concentrations have gotten worse. Recently, researchers estimated that by 2050 as many as 150 million people in the developing world will be at risk of protein deficiency as the result of nutrient collapse, since so many of the world's poor depend on crops, rather than animal meat, for protein; 138 million could suffer from a deficiency of zinc, essential to healthy pregnancies; and 1.4 billion could face a dramatic decline in dietary iron—pointing to a possible epidemic of anemia. In 2018, a team led by Chunwu Zhu looked at the protein content of eighteen different strains of rice, the staple crop for more than

2 billion people, and found that more carbon dioxide in the air produced nutritional declines across the board—drops in protein content, as well as in iron, zinc, and vitamins B_1, B_2, B_5, and B_9. Really everything but vitamin E. Overall, the researchers found that, acting just through that single crop, rice, carbon emissions could imperil the health of 600 million people.

In previous centuries, empires were built on that crop. Climate change promises another, an empire of hunger, erected among the world's poor.

Drowning

THAT THE SEA WILL BECOME A KILLER IS A GIVEN. BARRING a reduction of emissions, we could see at least four feet of sea-level rise and possibly eight by the end of the century. A radical reduction—of the scale that could make the Paris two-degree goal a conceivably attainable if quite optimistic target—could still produce as much as two meters, or six feet, by 2100.

Perversely, for a generation now, we've been comforted by numbers like these—when we think the worst that climate change can bring is an ocean a few feet higher, anyone who lives even a short distance from the coast feels like they can breathe easy. In that way, even alarmist popular writing about global warming has been a victim of its own success, so focused on sea-level rise that it has blinded readers to all the climate scourges beyond the oceans that threaten to terrorize the coming generations—direct heat, extreme weather, pandemic disease, and more. But as "familiar" as sea-level rise may seem, it surely deserves its place at the center of the picture of what damage climate change will bring. That so many feel already acclimated to the prospect of a near-future world with dramatically higher oceans should be as dispiriting and disconcerting as if we'd already come to accept the inevitability of extended nuclear war—because that is the scale of devastation the rising oceans will unleash.

In *The Water Will Come*, Jeff Goodell runs through just a few of the monuments—indeed, in some cases, whole cultures—that will be transformed into underwater relics, like sunken ships, this century: any beach you've ever visited; Facebook's headquarters, the Kennedy Space Center, and the United States' largest naval base, in Norfolk, Virginia; the entire nations of the Maldives and the Marshall Islands; most of Bangladesh, including all of the mangrove forests that have been the kingdom of Bengal tigers for millennia; all of Miami Beach and much of the South Florida paradise engineered out of marsh and swamp and sandbar by rabid real-estate speculators less than a century ago; Saint Mark's Basilica in Venice, today nearly a thousand years old; Venice Beach and Santa Monica in Los Angeles; the White House at 1600 Pennsylvania Avenue, as well as Trump's "Winter White House" at Mar-a-Lago, Richard Nixon's in Key Biscayne, and the original, Harry Truman's, in Key West. This is a very partial list. We've spent the millennia since Plato enamored with the story of a single drowned culture, Atlantis, which if it ever existed was probably a small archipelago of Mediterranean islands with a population numbering in the thousands—possibly tens of thousands. By 2100, if we do not halt emissions, as much as 5 percent of the world's population will be flooded every single year. Jakarta is one of the world's fastest-growing cities, today home to ten million; thanks to flooding and literal sinking, it could be entirely underwater as soon as 2050. Already, China is evacuating hundreds of thousands every summer to keep them out of the range of flooding in the Pearl River Delta.

What would be submerged by these floods are not just the homes of those who flee—hundreds of millions of new climate refugees unleashed onto a world incapable, at this point, of accommodating the needs of just a few million—but communities, schools, shopping districts, farmlands, office buildings and high-rises, regional cultures so sprawling that just a few centuries ago we might have remembered them as empires unto themselves, now suddenly underwater museums showcasing the way of life in the one or two centuries when humans, rather than keeping their safe distance, rushed to build up at the coastline. It will take thousands of years, perhaps millions, for quartz and feldspar to degrade into sand that might replenish the beaches we lose.

Much of the infrastructure of the internet, one study showed, could be drowned by sea-level rise in less than two decades; and most of the smartphones we use to navigate it are today manufactured in Shenzhen, which, sitting right in the Pearl River Delta, is likely to be flooded soon, as well. In 2018, the Union of Concerned Scientists found that nearly 311,000 homes in the United States would be at risk of chronic inundation by 2045—a timespan, as they pointed out, no longer than a mortgage. By 2100, the number would be more than 2.4 million properties, or $1 trillion worth of American real estate—underwater. Climate change may not only make the miles along the American coast uninsurable, it could render obsolete the very idea of disaster insurance; by the end of the century, one recent studied showed, certain places could be struck by six different climate-driven disasters simultaneously. If no significant action is taken to curb emissions, one estimate of global damages is as high as $100 trillion *per year* by 2100. That is more than global GDP today. Most estimates are a bit lower: $14 trillion a year, still almost a fifth of present-day GDP.

But the flooding wouldn't stop at the end of the century, since sea-level rise would continue for millennia, ultimately producing, in even that optimistic two-degree scenario, oceans six meters higher. What would that look like? The planet would lose about 444,000 square miles of land, where about 375 million people live today—a quarter of them in China. In fact, the twenty cities most affected by such sea-level rise are all Asian megalopolises—among them Shanghai, Hong Kong, Mumbai, and Kolkata. Which does cast a climate shroud over the prospect, now so much taken for granted among the Nostradamuses of geopolitics, of an Asian century. Whatever the course of climate change, China will surely continue its ascent, but it will do so while fighting back the ocean, as well—perhaps one reason it is already so focused on establishing control over the South China Sea.

Nearly two-thirds of the world's major cities are on the coast—not to mention its power plants, ports, navy bases, farmlands, fisheries, river deltas, marshlands, and rice paddies—and even those above ten feet will flood much more easily, and much more regularly, if the water gets that high. Already, flooding has quadrupled since 1980, according to the European Academies' Science Advisory Council, and doubled since just 2004. Even under an "intermediate low" sea-level-rise scenario, by 2100 high-tide flooding could hit the East Coast of the United States "every other day."

We haven't even gotten to inland flooding—when rivers run over, swollen by deluges of rain or storm surges channeled downstream from the sea. Between 1995 and 2015, this affected 2.3 billion and killed 157,000 around the world. Under even the most radically aggressive global emissions reduction regime, the further warming of the planet from just the carbon we've already pumped into the atmosphere would increase global rainfall to such a degree that the number affected by river flooding in South America would double, according to one paper, from 6 million to 12 million; in Africa, it would grow from 24 to 35 million, and in Asia from 70 to 156 million. All told, at just 1.5 degrees Celsius of warming, flood damage would increase by between 160 and 240 percent; at 2 degrees, the death toll from flooding would be 50 percent higher than today. In the United States, one recent model suggested that FEMA's recent projections of flood risk were off by a factor of three, and that more than 40 million Americans were at risk of catastrophic inundation.

These effects will come to pass even with a radical reduction of emissions, keep in mind. Without flood adaptation measures, large swaths of northern Europe and the whole eastern half of the United States will be affected by at least ten times as many floods. In large parts of India, Bangladesh, and Southeast Asia, where flooding is today catastrophically common, the multiplier could be just as high—and the baseline is already so elevated that it annually produces humanitarian crises on a scale we like to think we would not forget for generations.

Instead, we forget them immediately. In 2017, floods in South Asia killed 1,200 people, leaving two thirds of Bangladesh underwater; António Guterres, the secretary-general of the United Nations, estimated that 41 million people had been affected. As with so much climate change data, those numbers can numb, but 41 million is as much as eight times the entire global population at the time of the Black Sea deluge 7,600 years ago—reputedly so dramatic and catastrophic a flood that it may have given rise to our Noah's Ark story. At the same time as the floods hit in 2017, almost 700,000 Rohingya refugees from Myanmar arrived in Bangladesh, most of them in a single settlement site that became, in months, more populous than Lyon, France's third biggest city, and was erected in the path of landslides just as the next monsoon season arrived.

To what degree we will be able to adapt to new coastlines is primarily a matter of just how fast the water rises. Our understanding of that timeline has been evolving disconcertingly fast. When the Paris Agreement was drafted, those writing it were sure that the Antarctic ice sheets would remain stable even as the planet warmed several degrees; their expectation was that oceans could rise, at most, only three feet by the end of the century. That was just in 2015. The same year, NASA found that this expectation was hopelessly complacent, suggesting three feet was not a maximum but in fact a minimum. In 2017, the National Oceanic and Atmospheric Administration (NOAA) suggested eight feet was possible—still just in this century. On the East Coast, scientists have already introduced a new term, "sunny day flooding"—when high tide alone, aided by no additional rainstorm, inundates a town.

In 2018, a major study found things accelerating faster still, with the melt rate of the Antarctic ice sheet tripling just in the past decade. From 1992 to 1997, the sheet lost, on average, 49 billion tons of ice each year; from 2012 to 2017, it was 219 billion. In 2016, climate scientist James Hansen had suggested sea level could rise several meters over fifty years, if ice melt doubled every decade; the new paper, keep in mind, registers a tripling, and in the space of just five years. Since the 1950s, the continent has lost 13,000 square miles from its ice shelf; experts say its ultimate fate will probably be determined by what human action is taken in just the next decade.

All climate change is governed by uncertainty, mostly the uncertainty of human action—what action will be taken, and when, to avert or forestall the dramatic transformation of life on the planet that will unfold in the absence of dramatic intervention. Each of our projections, from the most blasé to the most extreme, comes wrapped in doubt—the result of so many estimates and so many assumptions that it would be foolish to take any of them, so to speak, to the bank.

But sea-level rise is different, because on top of the basic mystery of human response it layers much more epistemological ignorance than governs any other aspect of climate change science, save perhaps the question of cloud formation. When water warms, it expands: this we know. But the breaking-up of ice represents almost an entirely new physics, never before observed in human history, and therefore only poorly understood.

There are now, thanks to rapid Arctic melt, papers devoted to what are called the "damage mechanics" of ice-shelf loss. But we do not yet well understand those dynamics, which will be one of the main drivers of sea-level rise, and so cannot yet make confident predictions about how quickly ice sheets will melt. And even though we now have a decent picture of the planet's climatological past, never in the earth's entire recorded history has there been warming at anything like this speed—by one estimate, around ten times faster than at any point in the last 66 million years. Every year, the average American emits enough carbon to melt 10,000 tons of ice in the Antarctic ice sheets—enough to add 10,000 cubic meters of water to the ocean. Every minute, each of us adds five gallons.

One study suggests that the Greenland ice sheet could reach a tipping point at just 1.2 degrees of global warming. (We are nearing that temperature level today, already at 1.1 degrees.) Melting that ice sheet alone would, over centuries, raise sea levels six meters, eventually drowning Miami and Manhattan and London and Shanghai and Bangkok and Mumbai. And while business-as-usual emissions trajectories warm the planet by just over 4 degrees by 2100, because temperature changes are unevenly distributed around the planet, they threaten to warm the Arctic by 13.

In 2014, we learned that the West Antarctic and Greenland ice sheets were even more vulnerable to melting than scientists anticipated—in fact, the West Antarctic sheet had already passed a tipping point of collapse, more than doubling its rate of ice loss in just five years. The same had happened in Greenland, where the ice sheet is now losing almost a billion tons of ice every single day. The two sheets contain enough ice to raise global sea levels ten to twenty feet—each. In 2017, it was revealed that two glaciers in the East Antarctic sheet were also losing ice at an alarming rate— eighteen billion tons of ice each year, enough to cover New Jersey in three feet of ice. If both glaciers go, scientists expect, ultimately, an additional 16 feet of water. In total, the two Antarctic ice sheets could raise sea level by 200 feet; in many parts of the world, the shoreline would move by many miles. The last time the earth was four degrees warmer, as Peter Brannen has written, there was no ice at either pole and sea level was 260 feet higher. There were palm trees in the Arctic. Better not to think what that means for life at the equator.

As with all else in climate, the melting of the planet's ice will not occur in a vacuum, and scientists do not yet fully understand exactly what cascading effects such collapses will trigger. One major concern is methane, particularly the methane that might be released by a melting Arctic, where permafrost contains up to 1.8 trillion tons of carbon, considerably more than is currently suspended in the earth's atmosphere. When it thaws, some of it will evaporate as methane, which is, depending on how you measure, at least several dozen times more powerful a greenhouse gas than carbon dioxide.

When I first began seriously researching climate change, the risk from a sudden release of methane from the Arctic permafrost was considered quite low—in fact so low that most scientists derided casual discussion of it as reckless fearmongering and deployed mockingly hyperbolic terms like "Arctic methane time bomb" and "burps of death" to describe what they saw as a climate risk not much worth worrying about in the near term. The news since has not been encouraging: one *Nature* paper found that the release of Arctic methane from permafrost lakes could be rapidly accelerated by bursts of what is called "abrupt thawing," already under way. Atmospheric methane levels have risen dramatically in recent years, confusing scientists unsure of their source; new research suggests the amount of gas being released by Arctic lakes could possibly double going forward. It's not clear whether this methane release is new or just that we finally began to pay attention to it. But while the consensus is still that a rapid, sudden release of methane is unlikely, the new research is a case study in why it is worthwhile to consider, and take seriously, such unlikely-but-possible climate risks. When you define anything outside a narrow band of likelihood as irresponsible to consider, or talk about, or plan for, even unspectacular new research findings can catch you flat-footed.

Today, all do agree that that permafrost is melting—the permafrost line having retreated eighty miles north in Canada over the last fifty years. The most recent IPCC assessment projects a loss of near-surface permafrost of between 37 and 81 percent by 2100, though most scientists still believe that carbon will be released slowly, and mostly as less-terrifying carbon dioxide. But as far back as 2011, NOAA and the National Snow and Ice Data Center predicted that thawing permafrost would flip the whole region from being what is called a carbon sink, which absorbs atmospheric carbon, to a carbon source, which releases carbon, as quickly as the 2020s. By 2100, the same study said, the Arctic will have released a hundred billion tons of carbon. That is the equivalent of half of all the carbon produced by humanity since industrialization began.

Remember, that is the Arctic feedback loop that does not much concern many climate scientists in the near term. The one that concerns them more, at present, is what is called the "albedo effect": ice is white and so reflects sunlight back into space rather than absorbing it; the less ice, the more sunlight is absorbed as global warming; and the total disappearance of that ice, Peter Wadhams has estimated, could mean a massive warming equivalent to the entire last twenty-five years of global carbon emissions. The last twenty-five years of emissions, keep in mind, is about half of the total that humanity has ever produced—a scale of carbon production that has pushed the planet from near-complete climate stability to the brink of chaos.

All of this is speculative. But our uncertainty over each of these dynamics—ice sheet collapse, Arctic methane, the albedo effect—clouds our understanding only of the pace of change, not its scale. In fact, we do know what the endgame for oceans looks like, just not how long it will take us to get there.

How much sea-level rise is that? The ocean chemist David Archer is the researcher who has focused perhaps most acutely on what he calls the "long thaw" impacts of global warming. It may take centuries, he says, even millennia, but he estimates that ultimately, even at just three degrees of warming, sea-level rise will be at least fifty meters—that is, fully one hundred times higher than Paris predicted for 2100. The U.S. Geological Survey puts the ultimate figure at eighty meters, or more than 260 feet.

The world would perhaps not be made literally unrecognizable by that flooding, but the distinction is ultimately semantic. Montreal would be almost entirely underwater, as would London. The United States is an unexceptional example: at just 170 feet, more than 97 percent of Florida would disappear, leaving only a few hills in the Panhandle; and just under 97 percent of Delaware would be submerged. Oceans would cover 80 percent of Louisiana, 70 percent of New Jersey, and half of South Carolina, Rhode Island, and Maryland. San Francisco and Sacramento would be underwater, as would New York City, Philadelphia, Providence, Houston, Seattle, and Virginia Beach, among dozens of other cities. In many places, the coast would retreat by as much as one hundred miles. Arkansas and Vermont, landlocked today, would become coastal.

The rest of the world may fare even worse. Manaus, the capital of the Brazilian Amazon, would not just be on the oceanfront, but underneath its waters, as would Buenos Aires and the biggest city in landlocked Paraguay, Asunción, now more than five hundred miles inland. In Europe, in addition to London, Dublin would be underwater, as would Brussels, Amsterdam, Copenhagen and Stockholm, Riga and Helsinki and Saint Petersburg. Istanbul would flood, and the Black Sea and the Mediterranean would join. In Asia, you could forget the coastline cities of Doha and Dubai and Karachi and Kolkata and Mumbai (to name just a few) and would be able to trace the trail of underwater metropolises from what is now close to desert, in Baghdad, all the way to Beijing, itself a hundred miles inland.

That 260-foot rise is, ultimately, the ceiling—but it is a pretty good bet we will get there eventually. Greenhouse gases simply work on too long a timescale to avoid it, though what kind of human civilization will be around to see that flooded planet is very much to be determined. Of course, the scariest variable is how quickly that flood will come. Perhaps it will be a thousand years, but perhaps much sooner. More than 600 million people live within thirty feet of sea level today.

Wildfire

THE TIME BETWEEN THANKSGIVING AND CHRISTMAS IS
meant to be, in Southern California, the start of rainy season.
Not in 2017. The Thomas Fire, the worst of those that roiled the
region that fall, grew 50,000 acres in one day, eventually burning 440 square miles and forcing the evacuations of more than
100,000 Californians. A week after it was sparked, it remained,
in the ominous semi-clinical language of wildfires, merely "15%
contained." For a poetic approximation, it was not a bad estimate
of how much of a handle we have on the forces of climate change
that unleashed the Thomas Fire and the many other environmental
calamities for which it was an apocalyptic harbinger. That is to say,
hardly any.

"The city burning is Los Angeles's deepest image of itself," Joan
Didion wrote in "Los Angeles Notebook," collected in *Slouching
Towards Bethlehem*, published in 1968. But the cultural impression
is apparently not all that deep, since the fires that broke out in the

fall of 2017 produced, in headlines and on television and via text
messages, an astonished refrain of the adjectives "unthinkable,"
"unprecedented," and "unimaginable." Didion was writing about
the fires that had swept through Malibu in 1956, Bel Air in 1961,
Santa Barbara in 1964, and Watts in 1965; she updated her list in
1989 with "Fire Season," in which she described the fires of 1968,
1970, 1975, 1978, 1979, 1980, and 1982: "Since 1919, when the
county began keeping records of its fires, some areas have burned
eight times."

The list of dates cautions, on the one hand, against wildfire
alarmism—against a sort of cartoonishly Californian environmental panic, in which all observers are all-consumed by the present
instance of disaster. But all fires are not equal. Five of the twenty
worst fires in California history hit the state in the fall of 2017, a
year in which over nine thousand separate ones broke out, burning
through more than 1,240,000 acres—nearly two thousand square
miles made soot.

That October, in Northern California, 172 fires broke out in
just two days—devastation so cruel and sweeping that two different accounts were published in two different local newspapers
of two different aging couples taking desperate cover in pools as
the fires swallowed their homes. One couple survived, emerging
after six excruciating hours to find their house transformed into
an ash monument; in the other account, it was only the husband
who emerged, his wife of fifty-five years having died in his arms.
As Americans traded horror stories in the aftermath of those fires,
they could be forgiven for mixing up the stories, or being confused;
that climate terror could be so general as to provide variations on
such a theme had seemed, as recently as that September, impossible
to believe.

The following year offered another variation. In the summer of
2018, the fires were fewer in number, totaling only six thousand.
But just one, made up of a whole network of fires together called
the Mendocino Complex, burned almost half a million acres alone.
In total, more than two thousand square miles in the state turned

to flame, and smoke blanketed almost half the country. Things were worse to the north, in British Columbia, where more than three million acres burned, producing smoke that would—if it followed the pattern of previous Canadian plumes—travel across the Atlantic to Europe. Then, in November, came the Woolsey Fire, which forced the evacuation of 170,000, and the Camp Fire, which was somehow worse, burning through more than 200 square miles and incinerating an entire town so quickly that the evacuees, 50,000 of them, found themselves sprinting past exploding cars, their sneakers melting to the asphalt as they ran. It was the deadliest fire in California history, a record that had been set almost a century before, by the Griffith Park Fire of 1933.

If these wildfires were not unprecedented, in California at least, what did we mean when we called them that? Like September 11, which followed several decades of morbid American fantasies about the World Trade Center, this new class of terror looked to a horrified public like a climate prophecy, made in fear, now made real.

That prophecy was threefold. First, the simple intuition of climate horrors—an especially biblical premonition when the plague is out-of-control fire, like a dust storm of flame. Second, the expanding reach of wildfires in particular, which now can feel, in much of the West, only a gust of bad wind away. But perhaps the most harrowing of the ways in which the fires seemed to confirm our cinematic nightmares was the third: that climate chaos could breach our most imperious fortresses—that is, our cities.

With Hurricanes Katrina, Sandy, Harvey, Irma, and Michael, Americans have gotten acquainted with the threat of flooding, but water is just the beginning. In the affluent cities of the West, even those conscious of environmental change have spent the last few decades walking our street grids and driving our highways, navigating our superabundant supermarkets and all-everywhere internet and believing that we had built our way out of nature. We have not. A paradise dreamscape erected in a barren desert, L.A. has always been an impossible city, as Mike Davis has so brilliantly written. The sight of flames straddling the eight-lane I-405 is a reminder that it is still impossible. In fact, getting more so. For a time, we had come to believe that civilization moved in the other direction—making the impossible first possible and then stable and routine. With climate change, we are moving instead toward nature, and chaos, into a new realm unbounded by the analogy of any human experience.

TWO BIG FORCES CONSPIRE TO PREVENT US FROM NORMALIZING fires like these, though neither is exactly a cause for celebration. The first is that extreme weather won't let us, since it won't stabilize—so that even within a decade, it's a fair bet that these fires, which now occupy the nightmares of every Californian, will be thought of as the "old normal." The good old days.

The second force is also contained in the story of the wildfires: the way that climate change is finally striking close to home. Some quite special homes. The California fires of 2017 burned the state's wine crop, blowtorched million-dollar vacation properties, and threatened both the Getty Museum and Rupert Murdoch's Bel-Air estate. There may not be two better symbols of the imperiousness of American money than those two structures. Nearby, the sunshiny children's fantasia of Disneyland was quickly canopied, as the fires began to encroach, by an eerily apocalyptic orange sky. On local golf courses, the West Coast's wealthy still showed up for their tee times, swinging their clubs just yards from blazing fires in photographs that could not have been more perfectly staged to skewer the country's indifferent plutocracy. The following year, Americans watched the Kardashians evacuate via Instagram stories, then read about the private firefighting forces they employed, the rest of the state reliant on conscripted convicts earning as little as a dollar a day.

By accidents of geography and by the force of its wealth, the United States has, to this point, been mostly protected from the devastation climate change has already visited on parts of the less-developed world—mostly. The fact that warming is now hitting our wealthiest citizens is not just an opportunity for ugly bursts of liberal schadenfreude; it is also a sign of just how hard, and how indiscriminately, it is hitting. All of a sudden, it's getting a lot harder to protect against what's coming.

What is coming? Much more fire, much more often, burning much more land. Over the last five decades, the wildfire season in the western United States has already grown by two and a half months; of the ten years with the most wildfire activity on record, nine have occurred since 2000. Globally, since just 1979, the season has grown by nearly 20 percent, and American wildfires now burn twice as much land as they did as recently as 1970. By 2050, destruction from wildfires is expected to double again, and in some places within the United States the area burned could grow fivefold. For every additional degree of global warming, it could quadruple. What this means is that at three degrees of warming, our likely benchmark for the end of the century, the United States might be dealing with sixteen times as much devastation from fire as we are today, when in a single year ten million acres were burned. At four degrees of warming, the fire season would be four times worse still. The California fire captain believes the term is already outdated: "We don't even call it fire season anymore," he said in 2017. "Take the 'season' out—it's year-round."

But wildfires are not an American affliction; they are a global pandemic. In icy Greenland, fires in 2017 appeared to burn ten times more area than in 2014; and in Sweden, in 2018, forests in the Arctic Circle went up in flames. Fires that far north may seem innocuous, relatively speaking, since there are not so many people up there. But they are increasing more rapidly than fires in lower latitudes, and they concern climate scientists greatly: the soot and ash they give off can land on and blacken ice sheets, which then absorb more of the sun's rays and melt more quickly. Another Arctic fire broke out on the Russia-Finland border in 2018, and smoke from Siberian fires that summer reached all the way to the mainland United States. That same month, the twenty-first century's second-deadliest wildfire had swept through the Greek seaside, killing ninety-nine. At one resort, dozens of guests tried to escape the flames by descending a narrow stone staircase into the Aegean, only to be engulfed along the way, dying literally in each other's arms.

The effects of these fires are not linear or neatly additive. It might be more accurate to say that they initiate a new set of biological cycles. Scientists warn that, even as California is baked into brush by a drier future, making inevitable more and more damaging fires, the probability of unprecedented-seeming rainfalls will grow, too—as much as a threefold increase of events like that which produced the state's Great Flood of 1862. And mudslides are among the clearest illustrations of what new horrors that heralds; in Santa Barbara that January, the town's low-lying homes were pounded by the mountains' detritus cascading down the hillside toward the ocean in an endless brown river. One father, in a panic, put his young children up on his kitchen's marble countertop, thinking it the strongest feature of the house, then watched as a rolling boulder smashed through the bedroom where the children had been just moments before. One kindergartner who didn't make it was found close to two miles from his home, in a gulley traced by train tracks close to the waterfront, having been carried there, presumably, on a continuous wave of mud. Two miles.

Each year, globally, between 260,000 and 600,000 people die from smoke from wildfires, and Canadian fires have been linked to spikes in hospitalizations as far away as the Eastern Seaboard of the United States. Drinking water in Colorado was damaged for years by the fallout from a single wildfire in 2002. In 2014, Canada's Northwest Territories were blanketed with wildfire smoke, producing a 42 percent spike in hospital visits for respiratory ailments and what one study called a "profound" negative effect on individual well-being. "One of the strongest emotions that people felt was isolation," the lead researcher later said. "There's a sense of not being able to get away. Where do you go? There's smoke everywhere."

WHEN TREES DIE—BY NATURAL PROCESSES, BY FIRE, AT THE hands of humans—they release into the atmosphere the carbon stored within them, sometimes for as long as centuries. In this way, they are like coal. Which is why the effect of wildfires on emissions is among the most feared climate feedback loops—that the world's forests, which have typically been carbon sinks, would become carbon sources, unleashing all that stored gas. The impact can be especially dramatic when the fires ravage forests arising out of peat. Peatland fires in Indonesia in 1997, for instance, released up to 2.6 billion tons of carbon—40 percent of the average annual global emissions level. And more burning only means more warming only means more burning. In California, a single wildfire can entirely eliminate the emissions gains made that year by all of the state's aggressive environmental policies. Fires of that scale happen now every year. In this way, they make a mockery of the technocratic, meliorist approach to emissions reduction. In the Amazon, which in 2010 suffered its second "hundred-year drought" in the space of five years, 100,000 fires were found to be burning in 2017.

At present, the trees of the Amazon take in a quarter of all the carbon absorbed by the planet's forests each year. But in 2018, Jair Bolsonaro was elected president of Brazil promising to open the rain forest to development—which is to say, deforestation. How much damage can one person do to the planet? A group of Brazilian scientists has estimated that between 2021 and 2030, Bolsonaro's deforestation would release the equivalent of 13.12 gigatons of carbon. Last year, the United States emitted about 5 gigatons. This means that this one policy would have between two and three times the annual carbon impact of the entire American economy, with all of its airplanes and automobiles and coal plants. The world's worst emitter, by far, is China; the country was responsible for 9.1 gigatons of emissions in 2017. This means Bolsonaro's policy is the equivalent of adding, if just for a year, a whole second China to the planet's fossil fuel problem—and, on top of that, a whole second United States.

Globally, deforestation accounts for about 12 percent of carbon emissions, and forest fires produce as much as 25 percent. The ability of forest soils to absorb methane has fallen by 77 percent in just three decades, and some of those studying the rate of tropical deforestation believe it could deliver an additional 1.5 degrees Celsius of global warming even if fossil fuel emissions immediately ceased.

Historically, the emissions rate from deforestation was even higher, with the clearing of woods and flattening of forests causing 30 percent of emissions from 1861 to 2000; until 1980, deforestation played a greater role in increases of hottest-day records than did direct greenhouse-gas emissions. There is a public health impact as well: every square kilometer of deforestation produces twenty-seven additional cases of malaria, thanks to what is called "vector proliferation"—when the trees are cleared out, the bugs move in.

This is not simply a wildfire phenomenon; each climate threat promises to trigger similarly brutal cycles. The fires should be terrorizing enough, but it is the cascading chaos that reveals the true cruelty of climate change—it can upend and turn violently against us everything we have ever thought to be stable. Homes become weapons, roads become death traps, air becomes poison. And the idyllic mountain vistas around which generations of entrepreneurs and speculators have assembled entire resort communities become, themselves, indiscriminate killers—and are made, with each successive destabilizing event, only more likely to kill again.

Disasters
No Longer Natural

H UMANS USED TO WATCH THE WEATHER TO PROPHESY THE future; going forward, we will see in its wrath the vengeance of the past. In a four-degree-warmer world, the earth's ecosystem will boil with so many natural disasters that we will just start calling them "weather": out-of-control typhoons and tornadoes and floods and droughts, the planet assaulted regularly with climate events that not so long ago destroyed whole civilizations. The strongest hurricanes will come more often, and we'll have to invent new categories with which to describe them; tornadoes will strike much more frequently, and their trails of destruction could grow longer and wider. Hail rocks will quadruple in size.

Early naturalists talked often about "deep time"—the perception they had, contemplating the grandeur of this valley or that rock basin, of the profound slowness of nature. But the perspective changes when history accelerates. What lies in store for us is more like what aboriginal Australians, talking with Victorian anthropologists, called "dreamtime," or "everywhen": the semi-mythical experience of encountering, in the present moment, an out-of-time past, when ancestors, heroes, and demigods crowded an epic stage. You can find it already by watching footage of an iceberg collapsing into the sea—a feeling of history happening all at once.

It is. The summer of 2017, in the Northern Hemisphere, brought unprecedented extreme weather: three major hurricanes arising in quick succession in the Atlantic; the epic "500,000-year" rainfall of Hurricane Harvey, dropping on Houston a million gallons of water for nearly every single person in the entire state of Texas; the wildfires of California, nine thousand of them burning through more than a million acres, and those in icy Greenland, ten times bigger than those in 2014; the floods of South Asia, clearing 45 million from their homes.

Then the record-breaking summer of 2018 made 2017 seem positively idyllic. It brought an unheard-of global heat wave, with temperatures hitting 108 in Los Angeles, 122 in Pakistan, and 124 in Algeria. In the world's oceans, six hurricanes and tropical storms appeared on the radars at once, including one, Typhoon Mangkhut, that hit the Philippines and then Hong Kong, killing nearly a hundred and wreaking a billion dollars in damages, and another, Hurricane Florence, which more than doubled the average annual rainfall in North Carolina, killing more than fifty and inflicting $17 billion worth of damage. There were wildfires in Sweden, all the way in the Arctic Circle, and across so much of the American West that half the continent was fighting through smoke, those fires ultimately burning close to 1.5 million acres. Parts of Yosemite National Park were closed, as were parts of Glacier National Park in Montana, where temperatures also topped 100. In 1850, the area had 150 glaciers; today, all but 26 are melted.

By 2040, the summer of 2018 will likely seem normal. But extreme weather is not a matter of "normal"; it is what roars back at us from the ever-worsening fringe of climate events. This is among the scariest features of rapid climate change: not that it changes the everyday experience of the world, though it does that, and dramatically; but that it makes once-unthinkable outlier events much more common, and ushers whole new categories of disaster into the realm of the possible. Already, storms have doubled since 1980, according to the European Academies' Science Advisory Council; and it is now estimated that New York City will suffer "500-year" floods once every twenty-five years. But sea-level rise is more dramatic elsewhere, which means that storm surges will be distributed unequally; in some places, storms on that scale will hit even more frequently. The result is a radically accelerated experience of extreme weather—what was once centuries' worth of natural disaster compressed into just a decade or two. In the case of Hawaii's East Island, which disappeared underwater during a single hurricane, into a day or two.

The climate effects on extreme precipitation events—often called deluges or even "rain bombs"—are even clearer than those on hurricanes, since the mechanism is about as straightforward as it gets: warmer air can hold more moisture than cooler air. Already, there are 40 percent more intense rainstorms in the United States than in the middle of the last century. In the Northeast, the figure is 71 percent. The very heaviest downfalls are today three-quarters heavier than they were in 1958, and only getting more so. The island of Kauai, in Hawaii, is one of the wettest places on Earth, and has in recent decades endured both tsunamis and hurricanes; when a climate-change-driven rain event hit in April 2018, it literally broke the rain gauges, and the National Weather Service had to offer a best-guess estimate: fifty inches of water in twenty-four hours.

When it comes to extreme weather, we are already living in unprecedented times. In America, the damages from quotidian thunderstorms—the unexceptional kind—have increased more than sevenfold since the 1980s. Power outages from storms have doubled just since 2003. When Hurricane Irma first emerged, it was with such intensity that some meteorologists proposed creating an entirely new category of hurricane for it—a Category 6. And then came Maria, rolling through the Caribbean and devastating a string of islands for the second time in a single week—two storms of such intensity that the islands might be prepared to endure them once a generation, or perhaps even less often. In Puerto Rico, Maria wiped out power and running water for much of the island for months, flooding its agricultural lands so fully that one farmer predicted the island wouldn't produce any food for the next year.

In its aftermath, Maria also showcased one of the uglier aspects of our climate blindness. Puerto Ricans are U.S. citizens, and live not far from the mainland on an island millions of Americans have visited personally. And yet when climate disaster struck there, we processed their suffering, perhaps out of psychological self-interest, as foreign and far away. Trump barely mentioned Puerto Rico in the week after Maria, and while that may not surprise, neither did the Sunday talk shows. By the weekend, a few days after the hurricane traversed the island, it was off the front page of *The New York Times* as well. When Trump's feud with the heroic mayor of San Juan and his problematic visit to the island—during which he tossed paper towels into a crowd without power or water like T-shirts at a Knicks game—made the hurricane a partisan issue, Americans did begin to focus on the destruction a bit more. But the attention paid remains trivial compared to the humanitarian toll—and when compared to the response to natural disasters that have recently hit the American mainland. "We're getting some intimations of how the ruling class intends to handle the accumulating disasters of the Anthropocene," as the cultural theorist McKenzie Wark, of the New School, wrote. "We're on our own."

And in the future, all that was once unprecedented becomes quickly routine. Remember Hurricane Sandy? By 2100, floods of that scale are expected as many as seventeen times more often in New York. Katrina-level hurricanes are expected to double in frequency. Looking globally, researchers have found an increase of 25 to 30 percent in Category 4 and 5 hurricanes for just one degree Celsius of global warming. Between just 2006 and 2013, the

Philippines were hit by seventy-five natural disasters; over the last four decades in Asia, typhoons have intensified by between 12 and 15 percent, and the proportion of Category 4 and 5 storms has doubled; in some areas, it has tripled. By 2070, Asian megacities could lose as much as $35 trillion in assets due to storms, up from just $3 trillion in 2005.

We are so far from investing in adequate defenses against these storms that we are still building out into their paths—as though we are homesteaders staking claim to land cleared each summer by tornadoes, committing ourselves blindly to generations being punished by natural disaster. In fact, it is worse than that, since paving over stretches of vulnerable coast, as we've done most conspicuously in Houston and New Orleans, stops up natural drainage systems with concrete that extends each epic flood. We tell ourselves we are "developing" the land—in some cases, fabricating it from marsh. What we are really building are bridges to our own suffering, since it's not just those new concrete communities built right into the floodplain that are vulnerable, but all those communities behind them, built on the expectation that the old swampy coastline could protect them. Which does call into question just what we mean, in the age of the Anthropocene, by the phrase "natural disaster."

Dreamtime weather won't stop at the shore, but will blanket the life of every human living on the planet, no matter how far from the coast. The warmer the Arctic, the more intense the blizzards in the northern latitudes—that's what's given the American Northeast 2010's "Snowpocalypse," 2014's "Snowmageddon," and 2016's "Snowzilla."

The inland effects of climate change are being felt in warmer seasons, too. In April 2011—just one month—758 tornadoes swept the American countryside. The previous April record had been 267, and the most for any previous month in recorded history was 542. The next month, there was another wave, including the tornado that killed 138 people in Joplin, Missouri. What's called America's "tornado alley" has moved five hundred miles in just thirty years, and while, technically, scientists aren't sure that climate change increases tornado formation, the paths of destruction tornadoes leave are getting longer, and they are getting wider; they arise from thunderstorms, which are increasing—the number of days on which they are possible growing as much as 40 percent by 2100, according to one assessment. The United States Geological Survey—not a notably alarmist corner of even the temperamentally conservative federal bureaucracy—recently "war-gamed" an extreme weather scenario they called "ARkStorm": winter storms strike California, producing flooding in the Central Valley three hundred miles long and twenty miles wide, and more destructive flooding in Los Angeles, Orange County, and the Bay Area up north, altogether forcing evacuation of more than a million Californians; wind speeds reach hurricane levels of 125 miles per hour in parts of the state, and at least 60 miles per hour throughout much of it; landslides cascade down from the Sierra Nevada mountains; and damage, all told, reaches $725 billion, nearly three times the estimate for a massive earthquake in the state, the much-feared "Big One."

In the past, even the recent past, disasters like these arrived with otherworldly force and incomprehensible moral logic. We could see them coming, on radar and by satellite, but could not interpret them—not legibly, not in ways that really made sense of them in relation to one another. Even atheists and agnostics might find themselves whispering the phrase "act of God" in the aftermath of a hurricane, or wildfire, or tornado, if only to express how inexplicable it felt to endure such suffering with no author behind it, no one to blame for it. Climate change will change this.

EVEN AS WE SETTLE INTO THINKING OF NATURAL DISASTERS AS a regular feature of our weather, the scope of devastation and horror they bring will not diminish. There are cascade effects here, too: ahead of Hurricane Harvey, the state of Texas cut off Houston's air-quality monitors, fearing they'd be damaged; immediately afterward, a cloud of "unbearable" smells began drifting out of the city's petrochemical plants. Ultimately, nearly half a billion gallons of industrial wastewater surged out of a single petrochemical plant into Galveston Bay. All told, that one storm produced more than a hundred "toxic releases," including 460,000 gallons of gasoline, 52,000 pounds of crude oil, and a massive, quarter-mile-wide discharge of hydrogen chloride, which, when it mixes with moisture, becomes hydrochloric acid, "which can burn, suffocate, and kill."

Down the coast in New Orleans, the storm hit was less direct, but there the city had already been knocked offline—without a full complement of drain pumps after an August 5 storm. When Katrina had hit New Orleans in 2005, it was not walloping a thriving city—the 2000 population of 480,000 had declined from a peak of over 600,000 in 1960. After the storm, it was as low as 230,000. Houston is a different case. One of the fastest-growing cities in the country in 2017—greater Houston even included the fastest-growing suburb in the country that year—it has more than five times as many residents as New Orleans. It's a tragic irony that many of those new arrivals who moved into the path of this storm over the last decades were brought there by the oil business, which has worked tirelessly to undermine public understanding of climate change and derail global attempts at reducing carbon emissions. One suspects this is not the last 500-year storm those workers will see before retirement—nor the last to be seen by the hundreds of oil rigs off the coast of Houston, or the thousand more bobbing now elsewhere off the Gulf Coast, until the toll of our emissions becomes so brutally clear that those rigs are finally retired.

The phrase "500-year storm" is also very helpful on the question of resilience. Even a devastated community, buckled in suffering, can endure a long period of recovery if it is wealthy and politically stable and needs to rebuild only once a century—perhaps even once every fifty years. But rebuilding for a decade in the wake of spectacular storms that hit once a decade, or once every two decades, is an entirely different matter, even for countries as rich as the United States and regions as well-off as greater Houston. New Orleans is still reeling from Katrina, a dozen years on, with the Lower Ninth Ward barely one-third as populated as it was before the storm. And it surely doesn't help that the entire coastline of Louisiana is being swallowed by the sea, with 2,000 square miles already gone. The state loses a football field of land every single hour. In the Florida Keys, 150 miles of road need to be raised to stay ahead of sea level, costing as much as $7 million each mile, or up to $1 billion, total. The county's 2018 road budget was $25 million.

For the world's poor, recovery from storms like Katrina and Irma and Harvey, hitting more and more often, is almost impossible. The best choice is often simply to leave. In the months after Hurricane Maria devastated Puerto Rico, thousands of islanders arrived in Florida, thinking it might be for good. Of course, that land is disappearing, too.

Freshwater Drain

SEVENTY-ONE PERCENT OF THE PLANET IS COVERED IN water. Barely more than 2 percent of that water is fresh, and only 1 percent of that water, at most, is accessible, with the rest trapped mostly in glaciers. Which means, in essence, as *National Geographic* has calculated, only 0.007 percent of the planet's water is available to fuel and feed its seven billion people.

Think of freshwater shortages and you probably feel an itch in your throat, but in fact hydration is just a sliver of what we need water for. Globally, between 70 and 80 percent of freshwater is used for food production and agriculture, with an additional 10 to 20 percent set aside for industry. And the crisis is not principally driven by climate change—that 0.007 percent should be, believe it or not, plenty, not just for the seven billion of us here but for as many as nine billion, perhaps even a bit more. Of course, we are likely heading north of nine billion this century, to a global population of at least ten and possibly twelve billion. As with food scarcity, much of the growth is expected in parts of the world already most strained by water shortage—in this case, urban Africa. In many African countries already, you are expected to get by on as little as twenty liters of water each day—less than half of what water organizations say is necessary for public health. As soon as 2030, global water demand is expected to outstrip supply by 40 percent.

Today, the crisis is political—which is to say, not inevitable or necessary or beyond our capacity to fix—and, therefore, functionally elective. That is one reason it is nevertheless harrowing as a climate parable: an abundant resource made scarce through governmental neglect and indifference, bad infrastructure and contamination, careless urbanization and development. There is no need for a water crisis, in other words, but we have one anyway, and aren't doing much to address it. Some cities lose more water to leaks than they deliver to homes: even in the United States, leaks and theft account for an estimated loss of 16 percent of freshwater; in Brazil, the estimate is 40 percent. In both cases, as everywhere, scarcity plays out so nakedly on a stage defined by have-and-have-not inequities that the resulting drama of resource competition can hardly be called, truly, a competition; the deck is so stacked that water shortage looks more like a tool of inequality. The global result is that as many as 2.1 billion people around the world do not have access to safe drinking water, and 4.5 billion don't have safely managed water for sanitation.

Like global warming, the water crisis is soluble, at present. But that 0.007 percent leaves an awfully thin margin, and climate change will cut into it. Half of the world's population depends on seasonal melt from high-elevation snow and ice, deposits that are dramatically threatened by warming. Even if we hit the Paris targets, the glaciers of the Himalayas will lose 40 percent of their ice by 2100, or possibly more, and there could be widespread water shortages in Peru and California, the result of glacier melt. At four degrees, the snow-capped Alps could look more like Morocco's Atlas Mountains, with 70 percent less snow by the end of the century. As soon as 2020, as many as 250 million Africans could face water shortages due to climate change; by the 2050s, the number could hit a billion people in Asia alone. By the same year, the World Bank found, freshwater availability in cities around the world could decline by as much as two-thirds. Overall, according to the United Nations, five billion people could have poor access to freshwater by 2050.

The United States won't be spared—boomtown Phoenix is, for instance, already in emergency planning mode, which should not surprise, given that even London is beginning to worry over water shortages. But given the reassurances of wealth—which can buy stopgap solutions and additional short-term supply—the United States will not be the worst hit. In India, already, 600 million face "high to extreme water stress," according to a 2018 government report, and 200,000 people die each year from lacking or contaminated water. By 2030, according to the same report, India will have only half the water it needs. In 1947, when the country was formed, per capita water availability in Pakistan stood at 5,000 cubic meters; today, thanks mostly to population growth, it is at 1,000; and soon continued growth and climate change will bring it down to 400.

In the last hundred years, many of the planet's largest lakes have begun drying up, from the Aral Sea in central Asia, which was once the world's fourth largest and which has lost more than 90 percent of its volume in recent decades, to Lake Mead, which supplies much of Las Vegas's water and has lost as much as 400 billion gallons in a single year. Lake Poopó, once Bolivia's second biggest, has completely disappeared; Iran's Lake Urmia has shrunk more than 80 percent in thirty years. Lake Chad has more or less evaporated entirely. Climate change is only one factor in this story, but its impact is not going to shrink over time.

What goes on within those lakes that survive is perhaps just as distressing. In China's Lake Tai, for instance, the blooming of warmwater-friendly bacteria in 2007 threatened the drinking water of two million people; the heating-up of East Africa's Lake

Tanganyika has imperiled the fish stock harvested and eaten by millions in four adjacent, hungry nations. Freshwater lakes, by the way, account for up to 16 percent of the world's natural methane emissions, and scientists estimate that climate-fueled aquatic plant growth could double those emissions over the next fifty years.

We're already racing, as a short-term fix for the world's drought boom, to drain underground water deposits known as aquifers, but those deposits took millions of years to accumulate and aren't coming back anytime soon. In the United States, aquifers already supply a fifth of our water needs; as Brian Clark Howard has noted, wells that used to draw water at 500 feet now require pumps at least twice as deep. The Colorado River Basin, which serves water to seven states, lost twelve cubic miles of groundwater between 2004 and 2013; the Ogallala Aquifer in part of the Texas Panhandle lost 15 feet in a decade, and is expected to drain by 70 percent over the next fifty years in Kansas. In the meantime, they're fracking in that drinking water. In India, in just the next two years, twenty-one cities could exhaust their groundwater supply.

THE FIRST DAY ZERO IN CAPE TOWN WAS IN MARCH 2018, the day when the city, a few months earlier and enduring its worst drought in decades, had predicted its taps would run proverbially dry.

Sitting in a living room in a modern apartment in an advanced metropolis somewhere in the developed world, this threat may seem hard to credit—so many cities looking nowadays like fantasies of endless and on-demand abundance for the world's wealthy. But of all urban entitlements, the casual expectation of never-ending drinking water is perhaps the most deeply delusional. It takes quite a lot to bring that water to your sink, your shower, and your toilet.

As climate crises so often do, in Cape Town the drought aggravated existing conflicts. In a memorable first-person account written at the time, Capetonian Adam Welz described the episode, which did end before the city went completely dry, as an operatic enactment of familiar local problems: mostly wealthy whites complaining that mostly poor blacks, many of whom receive a small allocation free, were draining the water supply; social media aflame with accusations of idle or indifferent black South Africans leaving water pipes running unattended and shantytown businesses running off stolen water. Black South Africans pointed the finger at suburban whites with pools and lawns, making hay over "orgies of flushing in the toilet stalls of upscale shopping malls." Conspiracy theories circulated involving federal indifference and withheld Israeli technology, and accusations of bad faith bounced from local authorities to national ones to meteorologists—altogether serving, as is almost always the case when communities must respond collectively to climate threats, as a buffet of excuses to not act. At the peak of the crisis, the mayor announced that nearly two-thirds of the city, 64 percent, were failing to abide by the city's new water restrictions, which aimed to limit water use to 23 gallons per person each day. The average American goes through four to five times that much; in arid Utah, founded on a Mormon prophecy predicting the arrival of an Eden in the desert, the average citizen goes through, each day, 248 gallons. In February, Cape Town halved the individual allotment to 13 gallons, and the army prepared to secure the city's water facilities.

But accusations of individual irresponsibility were a kind of weaponized red herring, as they often are in communities reckoning with the onset of climate pain. We frequently choose to obsess over personal consumption, in part because it is within our control and in part as a very contemporary form of virtue signaling. But ultimately those choices are, in almost all cases, trivial contributors, ones that blind us to the more important forces. When it comes to freshwater, the bigger picture is this: personal consumption amounts to such a thin sliver that only in the most extreme droughts can it even make a difference. Even before the drought, one estimate found that South Africa had nine million people without any access to water for personal consumption at all; the amount of water required to satisfy the needs of those millions is only about one-third the amount of water used, each year, to produce the nation's wine crop. In California, where droughts are punctuated by outrage over pools and ever-green lawns, total urban consumption accounts for only 10 percent.

In South Africa, eventually, the crisis passed—a combination of aggressive water rationing and the end of the dry season. But you could be forgiven, considering the news coverage of Cape Town, for thinking that the South African city was the first to stare down a Day Zero. In fact, São Paulo did it in 2015, after a two-year drought, limiting water use to twelve hours a day for some residents in an aggressive rationing system that shuttered businesses and forced mass layoffs. In 2008, Barcelona, facing the worst drought the city had seen since Catalonians began keeping records, had to barge in drinking water from France. In southern Australia, the "millennium drought" began with low rainfall in 1996 and con-tinued, through a Death Valley–like trough that lasted eight years, beginning in 2001 and ending only when La Niña rainfall finally relieved the area in 2010. Rice and cotton production in the region fell 99 and 84 percent, respectively. Rivers and lakes shriveled up and wetlands turned acidic. In 2018, in the Indian city of Shimla, once the summertime home of the British Raj, the taps ran dry for weeks in May and June.

And while agriculture is often hit the hardest by shortages, water issues are not exclusively rural. Fourteen of the world's twenty biggest cities are currently experiencing water scarcity or drought. Four billion people, it is estimated, already live in regions facing water shortages at least one month each year—that's about two-thirds of the planet's population. Half a billion are in places where the shortages never end. Today, at just one degree of warm-ing, those regions with at least a month of water shortages each year include just about all of the United States west of Texas, where lakes and aquifers are being drained to meet demand, and stretch-ing up into western Canada and down to Mexico City; almost all of North Africa and the Middle East; a large chunk of India; almost all of Australia; significant parts of Argentina and Chile; and everything in Africa south of Zambia.

As long as it has had advocates, climate change has been sold under a saltwater banner—melting Arctic, rising seas, shrinking coastlines. A freshwater crisis is more alarming, since we depend on it far more acutely. It is also closer at hand. But while the planet commands the necessary resources today to provide water for drinking and sanitation to all the world's people, there is not the necessary political will—or even the inclination—to do so.

Over the next three decades, water demand from the global food system is expected to increase by about 50 percent, from cities and industry by 50 to 70 percent, and from energy by 85 percent. And climate change, with its coming megadroughts, promises to tighten supply considerably. In fact, the World Bank, in its landmark study of water and climate change "High and Dry," found that "the impacts of climate change will be channeled primarily through the water cycle." The bank's foreboding warning: that when it comes to the cruelly cascading effects of climate change, water efficiency is as pressing a problem, and as important a puzzle to solve, as energy efficiency. Without any meaningful adaptation in the distribution of water resources, the World Bank estimates, regional GDP could decline, simply due to water insecurity, by as much as 14 percent in the Middle East, 12 percent in Africa's Sahel, 11 percent in central Asia, and 7 percent in east Asia.

But of course GDP is at best a crude measure of environmental cost. A more eye-opening ledger is kept by Peter Gleick of the Pacific Institute: a simple list of all armed conflicts tied up with water issues, beginning in 3,000 BC with the ancient Sumerian legend of Ea. Gleick lists nearly five hundred water-related conflicts since 1900; almost half of the entire list is since just 2010. Part of that, Gleick acknowledges, is a reflection of the relative abundance of recent data, and part of it is the changing nature of war—conflicts that used to unfold almost exclusively between states are now, in an era where state authority has weakened in many places, likely to spark within states and between groups. The five-year Syrian drought that stretched from 2006 to 2011, producing crop failures that created political instability and helped usher in the civil war that produced a global refugee crisis, is one vivid example. Gleick is personally more focused on the strange war unfolding in Yemen since 2015—technically a civil war, but practically a proxy regional war between Saudi Arabia and Iran, and conceptually a sort of world war in miniature, with American and Russian involvement as well. There, the humanitarian cost has been carried as much by water as by blood; in part because of targeted attacks on water infrastructure, the number of cholera cases grew to one million in 2017, which means in a single year roughly 4 percent of the country contracted the disease.

"There's a saying in the water community," Gleick tells me. "If climate change is a shark, the water resources are the teeth."

Dying Oceans

W<small>E TEND TO SEE OCEANS AS UNFATHOMABLE, THE CLOS-</small>est thing we have on this planet to outer space: dark, forbidding, and, especially in the depths, quite weird and mysterious. "Who has known the ocean?" Rachel Carson wrote in her essay "Undersea," published twenty-five years before she tackled the desecration of the planet's land by human hands, and industrial "cure-alls," in *Silent Spring:* "Neither you nor I, with our earth-bound senses, know the foam and surge of the tide that beats over the crab hiding under the seaweed of his tide-pool home; or the lilt of the long, slow swells of mid-ocean, where shoals of wandering fish prey and are preyed upon, and the dolphin breaks the waves to breathe the upper atmosphere."

But the ocean isn't the other; we are. Water is not a beachside attraction for land animals: at 70 percent of the earth's surface it is, by an enormous margin, the planet's predominant environment. Along with everything else it does, oceans feed us: globally, seafood accounts for nearly a fifth of all animal protein in the human diet, and in coastal areas it can provide much more. The oceans also maintain our planetary seasons, through prehistoric currents like the Gulf Stream, and modulate the temperature of the planet, absorbing much of the heat of the sun.

Perhaps "has fed," "has maintained," and "has modulated" are better terms, since global warming threatens to undermine each of those functions. Already, fish populations have migrated north by hundreds of miles in search of colder waters—flounder by 250 miles off the American East Coast, mackerel so far from their Continental home that the fishermen chasing them are no longer bound by rules set by the European Union. One study tracing human impact on marine life found only 13 percent of the ocean undamaged, and parts of the Arctic have been so transformed by warming that scientists are beginning to wonder how long they can keep calling those waters "arctic." And however much sea-level rise and coastal flooding have dominated our fears about the impact of climate change on the planet's ocean water, there is much more than just that to worry over.

At present, more than a fourth of the carbon emitted by humans is sucked up by the oceans, which also, in the past fifty years, have absorbed 90 percent of global warming's excess heat. Half of that heat has been absorbed since 1997, and today's seas carry at least 15 percent more heat energy than they did in the year 2000—absorbing three times as much additional energy, in just those two decades, as is contained in the entire planet's fossil fuel reserves. But the result of all that carbon dioxide absorption is what's called "ocean acidification," which is exactly what it sounds like, and which is also already burning through some of the planet's water basins—you may remember these as the place where life arose in the first place. All on its own—through its effect on phytoplankton, which release sulfur into the air that helps cloud formation—ocean acidification could add between a quarter and half of a degree of warming.

YOU HAVE PROBABLY HEARD OF "CORAL BLEACHING"—THAT IS, coral dying—in which warmer ocean waters strip reefs of the protozoa, called zooxanthellae, that provide, through photosynthesis, up to 90 percent of the energy needs of the coral. Each reef is an ecosystem as complex as a modern city, and the zooxanthellae are its food supply, the basic building block of an energy chain; when they die, the whole complex is starved with military efficiency, a city under siege or blockade. Since 2016, as much as half of Australia's landmark Great Barrier Reef has been stripped in this way. These large-scale die-outs are called "mass bleaching events"; one unfolded, globally, from 2014 to 2017. Already, coral life has declined so much that it has created an entirely new layer in the ocean, between 30 and 150 meters below the surface, which scientists have taken to calling a "twilight zone." According to the World Resources Institute, by 2030 ocean warming and acidification will threaten 90 percent of all reefs.

This is very bad news, because reefs support as much as a quarter of all marine life and supply food and income for half a billion people. They also protect against flooding from storm surges—a function that offers value in the many billions, with reefs presently worth at least $400 million annually to Indonesia, the Philippines, Malaysia, Cuba, and Mexico—$400 million annually to each. Ocean acidification will also damage fish populations directly. Though scientists aren't yet sure how to predict the effects on the stuff we haul out of the ocean to eat, they do know that in acid waters, oysters and mussels will struggle to grow their shells, and that rising carbon concentrations will impair fishes' sense of smell—which you may not have known they had, but which often aids in navigation. Off the coasts of Australia, fish populations have declined an estimated 32 percent in just ten years.

It has become quite common to say that we are living through a mass extinction—a period in which human activity has multiplied the rate at which species are disappearing from the earth by a factor perhaps as large as a thousand. It is probably also fair to call this an era marked by what is called ocean anoxification. Over the past fifty years, the amount of ocean water with no oxygen at all has quadrupled globally, giving us a total of more than four hundred "dead zones"; oxygen-deprived zones have grown by several million square kilometers, roughly the size of all of Europe; and hundreds of coastal cities now sit on fetid, under-oxygenated ocean. This is partly due to the simple warming of the planet, since warmer waters can carry less oxygen. But it is also partly the result of straightforward pollution—a recent Gulf of Mexico dead zone, all 9,000 square miles of it, was powered by the runoff of fertilizer chemicals washing into the Mississippi from the industrial farms of the Midwest. In 2014, a not-atypical toxic event struck Lake Erie, when fertilizer from corn and soy farms in Ohio spawned an algae bloom that cut off drinking water for Toledo. And in 2018, a dead zone the size of Florida was discovered in the Arabian Sea—so big that researchers believed it might encompass the entire 63,700-square-mile Gulf of Oman, seven times the size of the dead zone in the Gulf of Mexico. "The ocean," said the lead researcher Bastien Queste, "is suffocating."

Dramatic declines in ocean oxygen have played a role in many of the planet's worst mass extinctions, and this process by which dead zones grow—choking off marine life and wiping out fisheries—is already quite advanced not only in the Gulf of Mexico but just off Namibia, where hydrogen sulfide is bubbling out of the sea along a thousand-mile stretch of land known as the Skeleton Coast. The name originally referred to the detritus of wrecked ships, but today it's more apt than ever. Hydrogen sulfide is also one of the things scientists suspect finally capped the end-Permian extinction, once all the feedback loops had been triggered. It is so toxic that evolution has trained us to recognize the tiniest, safest traces, which is why our noses are so exquisitely skilled at registering flatulence.

AND THEN THERE IS THE POSSIBLE SLOWDOWN OF THE "OCEAN conveyor belt," the great circulatory system made up of the Gulf Stream and other currents that is the primary way the planet regulates regional temperatures. How does this work? The water of the Gulf Stream cools off in the atmosphere of the Norwegian Sea, making the water itself denser, which sends it down into the bottom of the ocean, where it is then pushed southward by more Gulf Stream water—itself cooling in the north and falling to the ocean floor—eventually all the way to Antarctica, where the cold water returns to the surface and begins to heat up and travel north. The trip can take a thousand years.

As soon as the conveyor belt became the subject of real study, in the 1980s, there were those oceanographers who worried it might shut down, which would lead to a dramatic disequilibration of the planet's climate—the hotter parts getting much hotter and the colder parts much colder. A total shutdown would be inconceivably catastrophic, though the impacts look deceptively innocuous on first scan—a colder Europe, more intense weather, additional sea-level rise. Invariably, this is described as the *Day After Tomorrow* scenario, and it is a strange twist of fate that so forgettable a movie has become the memorable shorthand for this particular worst-case nightmare.

A shutdown of the conveyor belt is not a scenario that any credible scientists worry about on any human timescale. But a slowdown is another matter. Already, climate change has depressed the velocity of the Gulf Stream by as much as 15 percent, a development that scientists call "an unprecedented event in the past millennium," believed to be one reason the sea-level rise along the East Coast of the United States is dramatically higher than elsewhere in the world. And in 2018, two major papers triggered a new wave of concern over the conveyor belt, technically called Atlantic Meridional Overturning Circulation, which was found to be moving at its slowest rate in at least 1,500 years. This had happened about a hundred years ahead of the schedule of even alarmed scientists and marked what the climate scientist Michael Mann called, ominously, a "tipping point." Further change, of course, is to come: the transformation of the ocean by warming making these unknown waters doubly unknowable, remodeling the planet's seas before we ever were able to discover their depths and all the life submerged there.

Unbreathable Air

O UR LUNGS NEED OXYGEN, BUT IT IS ONLY A FRACTION OF what we breathe, and the fraction tends to decline the more carbon is in the atmosphere. That doesn't mean we are at risk of suffocation—oxygen is far too abundant for that—but we will nevertheless suffer. With CO_2 at 930 parts per million (more than double where we are today), cognitive ability declines by 21 percent.

The effects are more pronounced indoors, where CO_2 tends to build up—that's one reason you probably feel a little more awake when taking a brisk walk outside than you do after spending a long day inside with the windows closed. And it's also a reason elementary school classrooms have been found, by one study, to already average 1,000 parts per million, with almost a quarter of those surveyed in Texas over 3,000—quite alarming numbers, given that these are the environments we've designed to promote intellectual performance. But classrooms are not the worst offenders: other studies have shown even higher concentrations on airplanes, with effects you can probably groggily recall from past experience.

But carbon is, more or less, the least of it. Going forward, the planet's air won't just be warmer; it will likely also be dirtier, more oppressive, and more sickening. Droughts have a direct impact on air quality, producing what is now known as dust exposure and in the days of the American Dust Bowl was called "dust pneumonia"; climate change will bring new dust storms to those plains states, where deaths from dust pollution are expected to more than double and hospitalizations to triple. The hotter the planet gets, the more ozone forms, and by the middle of this century Americans should suffer a 70 percent increase in days with unhealthy ozone smog, the National Center for Atmospheric Research has projected. By the 2090s, as many as 2 billion people globally will be breathing air above the WHO "safe" level. Already, more than 10,000 people die from air pollution daily. That is considerably more each day—*each day*—than the total number of people who have ever been affected by the meltdowns of nuclear reactors. This is not a slam-dunk argument in favor of nuclear power, of course, since the comparison isn't so neat: there are many, many more fossil fuel chimneys disgorging their trails of black smoke than fission facilities with their finger-trap towers and clouds of white vapor. But it is a startling mark of just how all-encompassing our regime of carbon pollution really is, enclosing the planet in a toxic swaddle.

In recent years, researchers have uncovered a whole secret history of adversity woven into the experience of the last half century by the hand of leaded gasoline and lead paint, which seem to have dramatically increased rates of intellectual disability and criminality, and dramatically decreased educational attainment and lifetime earnings, wherever they were introduced. The effects of air pollution seem starker already. Small-particulate pollution, for instance, lowers cognitive performance over time so much that researchers call the effect "huge": reducing Chinese pollution to the EPA standard, for instance, would improve the country's verbal test scores by 13 percent and its math scores by 8 percent. (Simple temperature rise has a robust and negative impact on test taking, too: scores go down when it's hotter out.) Pollution has been linked with increased mental illness in children and the likelihood of dementia in adults. A higher pollution level in the year a baby is born has been shown to reduce earnings and labor force participation at age thirty, and the relationship of pollution to premature births and low birth weight of babies is so strong that the simple introduction of E-ZPass in American cities reduced both problems, in the vicinity of toll plazas, by 10.8 percent and 11.8 percent, respectively, just by cutting down on the exhaust expelled when cars slowed to pay the toll.

Then there is the more familiar health threat from pollution. In 2013, melting Arctic ice remodeled Asian weather patterns, depriving industrial China of the natural wind-ventilation patterns it had come to depend on, and, as a result, blanketing much of the country's north in an unbreathable smog. An obtuse-seeming metric called the Air Quality Index categorizes the risks according to an idiosyncratic unit scale tabulating the presence of a variety of pollutants: the warnings begin at 51–100, and at 201–300 include promises of "significant increase in respiratory effects in the general population." The index tops out with the 301–500 range, warning of "serious aggravation of heart or lung disease and premature mortality in persons with cardiopulmonary disease and the elderly" and "serious risk of respiratory effects in the general population"; at that level, "everyone should avoid all outdoor exertion." The Chinese "airpocalypse" of 2013 doubled the high end of that upper range, reaching a peak Air Quality Index of 993, and scientists studying the phenomenon suggested that China had inadvertently invented an entirely new and unstudied kind of smog, one that combined the "pea soup" pollution of industrial-era Europe and the small-particulate pollution that has lately contaminated so much of the developing world. That year, smog was responsible for 1.37 million deaths in the country.

Outside of China, most saw the photographs and video of a

world capital blanketed by gray so thick it blotted out the sun as a sign, not of the state of the planet's atmosphere, but of just how backward that one country was—just how far China lagged behind the quality-of-life indices of the first world, whatever its rapid economic growth suggested about its place in the global pecking order. Then, in the record California wildfire season of 2017, the air around San Francisco was worse than on the same day in Beijing. In Napa, the Air Quality Index hit 486. In Los Angeles, there was a run on surgical masks; in Santa Barbara, residents scooped ash from their drainpipes by the handful. In Seattle, the following year, wildfire smoke made it unsafe for anyone, anywhere, to breathe outside. Which gave Americans one more reason—panic about their own health—to look away from the situation in Delhi, where in 2017 the Air Quality Index reached 999.

The Indian capital is home to 26 million people. In 2017, simply breathing its air was the equivalent of smoking more than two packs of cigarettes a day, and local hospitals saw a patient surge of 20 percent. Runners in Delhi's half marathon competed with their heads wrapped by white masks. And air that thick with smut is hazardous in other ways: visibility was so low that cars crashed in pileups on Delhi's highways, and United canceled flights in and out of the city.

New research shows that even short-term exposure to particulate pollution can dramatically increase rates of respiratory infections, with every additional ten micrograms per cubic meter associated with a rise in diagnoses between 15 and 32 percent. Blood pressure goes up, too. In 2017, *The Lancet* reported, nine million premature deaths globally were from small-particulate pollution; more than a quarter were in India. And that was before final figures were in from that year's spike.

In Delhi, much of the pollution comes from the burning of nearby farmland; but elsewhere small-particulate smog is produced primarily by diesel and gas exhaust and other industrial activity. The public health damage is indiscriminate, touching nearly every human vulnerability: pollution increases prevalence of stroke, heart disease, cancer of all kinds, acute and chronic respiratory diseases like asthma, and adverse pregnancy outcomes, including premature birth. New research into the behavioral and developmental effects is perhaps even scarier: air pollution has been linked to worse memory, attention, and vocabulary, and to ADHD and autism spectrum disorders. Pollution has been shown to damage the development of neurons in the brain, and proximity to a coal plant can deform your DNA.

In the developing world, 98 percent of cities are enveloped by air above the threshold of safety established by the WHO. Get out of urban areas and the problem doesn't much improve: 95 percent of the world's population is breathing dangerously polluted air. Since 2013, China has undertaken an unprecedented cleanup of its air, but as of 2015 pollution was still killing more than a million Chinese each year. Globally, one out of six deaths is caused by air pollution.

POLLUTION LIKE THIS ISN'T NEWS IN ANY MEANINGFUL SENSE; you can find omens about the toxicity of smog and the dangers of blackened air, for instance, in the writing of Charles Dickens, rarely appreciated as an environmentalist. But every year we are discovering more and more ways in which our industrial activity is poisoning the planet.

One particular note of alarm has been struck by what seems like an entirely new—or newly understood—pollution threat: microplastics. Global warming did not bring us microplastics in any direct way, and yet their rapid conquest of our natural world has become an irresistible fable about just what kind of transformation is meant by the word "Anthropocene," and just how much the world's booming consumer culture is to blame.

Environmentalists probably know already about "the Great Pacific garbage patch"—that mass of plastic, twice the size of Texas, floating freely in the Pacific Ocean. It is not actually an island—in fact, it is not actually a stable mass, only rhetorically convenient for us to think of it that way. And it is mostly composed of larger-scale plastics, of the kind visible to the human eye. The microscopic bits—700,000 of them can be released into the surrounding environment by a single washing-machine cycle—are more insidious. And, believe it or not, more pervasive: a quarter of fish sold in Indonesia and California contain plastics, according to one recent study. European eaters of shellfish, one estimate has suggested, consume at least 11,000 bits each year.

The direct effect on ocean life is even more striking. The total number of marine species said to be adversely affected by plastic pollution has risen from 260 in 1995, when the first assessment was carried out, to 690 in 2015 and 1,450 in 2018. A majority of fish tested in the Great Lakes contained microplastics, as did the guts of 73 percent of fish surveyed in the northwest Atlantic. One U.K. supermarket study found that every 100 grams of mussels were infested with 70 particles of plastic. Some fish have learned to eat plastic, and certain species of krill are now functioning as plastic processing plants, churning microplastics into smaller bits that scientists are now calling "nanoplastics." But krill can't grind it all down; in one square mile of water near Toronto, 3.4 million microplastic particles were recently trawled. Of course, seabirds are not immune: one researcher found 225 pieces of plastic in the stomach of a single three-month-old chick, weighing 10 percent of its body mass—the equivalent of an average human carrying about ten to twenty pounds of plastic in a distended belly. ("Imagine having to take your first flight out to sea with all that in your stomach," the researcher told the *Financial Times*, adding: "Around the world, seabirds are declining faster than any other bird group.")

Microplastics have been found in beer, honey, and sixteen of seventeen tested brands of commercial sea salt, across eight different countries. The more we test, the more we find; and while nobody yet knows the health impact on humans, in the oceans a plastic microbead is said to be one million times more toxic than the water around it. Chances are, if we started slicing open human cadavers to look for microplastics—as we are beginning to do with tau proteins, the supposed markers of CTE and Alzheimer's—

we'd be finding plastic in our own flesh, too. We can breathe in microplastics, even when indoors, where they've been detected suspended in the air, and do already drink them: they are found in the tap water of 94 percent of all tested American cities. And global plastic production is expected to triple by 2050, when there will be more plastic in the ocean than fish.

PLASTIC PANIC HAS A STRANGE RELATIONSHIP TO CLIMATE change, in that it seems to draw on premonitions about the degradation of the planet while focusing on something that has very little to do with global warming. But it's not only carbon emissions that are tied up in climate change. Other pollution is, too. One of the connections is relatively attenuated: plastics are produced by industrial activity that also produces pollutants, including carbon dioxide. A second is more direct but, in the scheme of things, trivial: when plastics degrade, they release methane and ethylene, another powerful greenhouse gas.

But a third relationship between non-carbon pollution and the temperature of the planet is far more horrifying. This is not the problem of plastic but of "aerosol pollution"—the blanket term for any particles suspended in our atmosphere. Aerosol particles actually suppress global temperature, mostly by reflecting sunlight back into outer space. In other words, all of the non-carbon pollution we've exhausted from our power plants and our factories and our automobiles—suffocating some of the largest and most prosperous cities of the world and consigning many millions of the lucky to hospital beds, and many millions of others to early deaths—all of that pollution has been, perversely, reducing the amount of global warming we are currently experiencing.

How much? Probably about half a degree—and possibly more. Already, aerosols have been reflecting so much sunlight away from the earth that, in the industrial era, the planet has only heated up two-thirds as much as it would have otherwise. If we had somehow managed to produce precisely the same volume of carbon emissions since the beginning of the Industrial Revolution as we have, while somehow keeping the skies clear of aerosol pollution, the temperature rise would be half again higher than it is now. The result is what the Nobel laureate Paul Crutzen has called a "Catch-22" and what the climate writer Eric Holthaus has described, perhaps more incisively, as a "devil's bargain": a choice between public-health-destroying pollution on the one hand, and, on the other, clear skies whose very clearness and healthiness will dramatically accelerate climate change. Eliminate that pollution and you save millions of lives each year, but also create a dramatic spike in warming. That would bring us to between 1.5 and 2 degrees warmer than the preindustrial baseline—pushing us right up to the threshold of 2 degrees of warming, long thought to be the border separating a livable future from climate catastrophe.

For almost a generation now, engineers and futurists have contemplated the practical implications of this phenomenon, and the prospect of suppressing global temperature with a program of suspended particles—that is, polluting the air on purpose to keep the planet cooler. Often grouped together under the umbrella term "geoengineering," this prospect has been received by the public as a worst-case scenario, nearly science fiction—and has, in fact, informed much of the recent sci-fi that has addressed itself to the climate crisis. And yet it has gained a terrific amount of currency among the most concerned climate scientists, many of whom will also note that none of the quite modest goals of the Paris climate accords can be achieved without negative-emissions technologies—at present prohibitively expensive.

Carbon capture may indeed prove to be "magical thinking," but the cruder technologies—we know these will work. Rather than sucking carbon out of the atmosphere, we could shoot pollution into the sky on purpose; perhaps the most plausible version involves sulfur dioxide. That would turn our sunsets very red, would bleach the sky, would make more acid rain.

It would also cause tens of thousands of additional premature deaths each year, through its effect on air quality. A 2018 paper suggested it would rapidly dry the Amazon, producing many more wildfires. The negative effect on plant growth would entirely cancel out the positive effect on global temperature, according to another 2018 paper; in other words, at least in terms of agricultural yield, solar geoengineering would offer no net benefit at all.

Once we began such a program, we could never stop. Even a brief interruption, a temporary dispersal of our red sulfur umbrella, could send the planet plunging several degrees of warming forward into a climate abyss. Which would make whatever installations were sustaining that umbrella quite vulnerable to political gamesmanship and terrorism, as its advocates themselves would acknowledge. And yet many scientists still describe geoengineering as an inevitability—it's just so cheap, they say. Even an environmentalist billionaire, going rogue, could make it happen on their own.

Plagues of
Warming

ROCK IS A RECORD OF PLANETARY HISTORY, ERAS AS LONG AS
millions of years flattened by the forces of geological time into
strata with amplitudes of just inches, or just an inch, or even less.
Ice works that way, too, as a climate ledger, but it is also frozen
history, some of which can be reanimated when unfrozen. There
are now, trapped in Arctic ice, diseases that have not circulated
in the air for millions of years—in some cases, since before humans were around to encounter them. Which means our immune
systems would have no idea how to fight back when those prehistoric plagues emerge from the ice. Already, in laboratories, several
microbes have been reanimated: a 32,000-year-old "extremophile"
bacteria revived in 2005, an 8-million-year-old bug brought back
to life in 2007, a 3.5-million-year-old one a Russian scientist self-
injected, out of curiosity, just to see what would happen. (He survived.) In 2018, scientists revived something a bit bigger—a worm
that had been frozen in permafrost for the last 42,000 years.

The Arctic also stores terrifying diseases from more recent times. In Alaska, researchers have discovered remnants of the 1918 flu that infected as many as 500 million, and killed as many as 50 million—about 3 percent of the world's population, and almost six times as many as had died in the world war for which the pandemic served as a kind of gruesome capstone. Scientists suspect smallpox and the bubonic plague are trapped in Siberian ice, among many other diseases that have otherwise passed into human legend—an abridged history of devastating sickness, left out like egg salad in the Arctic sun.

Many of these frozen organisms won't actually survive the thaw; those that have been brought back to life have been reanimated typically under fastidious lab conditions. But in 2016, a boy was killed and twenty others infected by anthrax released when retreating permafrost exposed the frozen carcass of a reindeer killed by the bacteria at least seventy-five years earlier; more than two thousand present-day reindeer died.

WHAT CONCERNS EPIDEMIOLOGISTS MORE THAN ANCIENT diseases are existing scourges relocated, rewired, or even re-evolved by warming. The first effect is geographical. Before the early modern period, human provinciality was a guard against pandemic—a bug could wipe out a town, or a kingdom, or even in an extreme case devastate a continent—but in most instances it couldn't travel much farther than its victims, which is to say, not very far at all. The Black Death killed as much as 60 percent of Europe, but consider, for a gruesome counterfactual, how big its impact might have been in a truly globalized world.

Today, even with globalization and the rapid intermingling of human populations, our ecosystems are mostly stable, and this functions as another limit—we know where certain bugs can spread, and know the environments in which they cannot. (This is why certain vectors of adventure tourism require dozens of new vac-

cines and prophylactic medications, and why New Yorkers travel-
ing to London don't need to worry.)

But global warming will scramble those ecosystems, meaning
it will help disease trespass those limits as surely as Cortés did.
The footprint of every mosquito-borne illness is presently circum-
scribed, but those borders are disappearing rapidly, as the tropics expand—the current rate is thirty miles per decade. In Brazil,
for generations, yellow fever sat in the Amazon basin, where the
Haemagogus and *Sabethes* mosquitoes thrived, making the disease
a concern for those who lived, worked, or traveled deep into the
jungle, but only for them; in 2016, it left the Amazon, as more and
more mosquitoes fanned out of the rain forest; and by 2017 it had
reached areas around the country's megalopolises, São Paulo and
Rio de Janeiro—more than thirty million people, many of them
living in shantytowns, facing the arrival of a disease that kills between 3 and 8 percent of those infected.

Yellow fever is just one of the plagues that will be carried by
mosquitoes as they migrate, conquering more and more of a warming world—the globalization of pandemic disease. Malaria alone
kills a million people each year already, infecting many more, but
you don't worry much about it if you are living in Maine or France.
As the tropics creep northward and mosquitoes migrate with them,
you may; over the course of the next century, more and more of the
world's population will be living under the shadow of diseases like
these. You didn't much worry about Zika before a couple of years
ago, either.

As it happens, Zika may also be a good model of a second
worrying effect—disease mutation. One reason you hadn't heard
about Zika until recently is that it had been trapped in Uganda and
Southeast Asia; another is that it did not, until recently, appear to
cause birth defects. Scientists still don't entirely understand what
happened or what they missed, even now, several years after the
planet seemed gripped by panic about microcephaly: it could be
that the disease changed as it came to the Americas, the result of

a genetic mutation or in adaptive response to a new environment; or that Zika produces those devastating prenatal effects only when another disease is present, possibly one less common in Africa; or that something about the environment or immunological history in Uganda protects mothers and their unborn children.

But there are things we do know for sure about how climate affects some diseases. Malaria, for instance, thrives in hotter regions, which is one reason the World Bank estimates that by 2030, 3.6 billion people will be reckoning with it—100 million as a direct result of climate change.

PROJECTIONS LIKE THOSE DEPEND NOT JUST ON CLIMATE MODels but on an intricate understanding of the organism at play. Or, rather, organisms. Malaria transmission involves both the disease and the mosquito; Lyme disease, both the disease and the tick—which is another epidemiologically threatening creature whose universe is rapidly expanding, thanks to global warming. As Mary Beth Pfeiffer has documented, Lyme case counts have spiked in Japan, Turkey, and South Korea, where the disease was literally nonexistent as recently as 2010—zero cases—and now lives inside hundreds more Koreans each year. In the Netherlands, 54 percent of the country's land is now infested; in Europe as a whole, Lyme caseloads are now three times the standard level. In the United States, there are likely around 300,000 new infections each year—and since many of even those treated for Lyme continue to show symptoms years after treatment, the numbers can stockpile. Overall, the number of disease cases from mosquitoes, ticks, and fleas have tripled in the U.S. over just the last thirteen years, with dozens of counties across the country encountering ticks for the first time. But the effects of the epidemic can be seen perhaps most clearly in animals other than humans: in Minnesota, during the 2000s, winter ticks helped drop the moose population by 58 percent in a single decade, and some environmentalists believe the species could be eradicated entirely from the state as soon as 2020.

In New England, dead moose calves have been found suckling as many as 90,000 engorged ticks, often killing the calves not through Lyme disease but simple anemia, the effect of that number of bugs each drawing a few milliliters of blood from the moose. Those that survive are far from robust, many having scratched so incessantly at their own hides to clear it of ticks that they completely eliminated their own hair, leaving behind a spooky gray skin that has earned them the name "ghost moose."

Lyme is still, in relative terms, a young disease, and one we don't yet understand all that well: we attribute a very mysterious and incoherent set of symptoms to it, from joint pain to fatigue to memory loss to facial palsy, almost as a catchall explanation for ailments we cannot pinpoint in patients who we know have been bitten by a bug carrying the bug. We do know ticks, however, as surely as we know malaria—there are not many parasites we understand better. But there are many, many millions we understand worse, which means our sense of how climate change will redirect or remodel them is shrouded in a foreboding ignorance. And then there are the plagues that climate change will confront us with for the very first time—a whole new universe of diseases humans have never before known to even worry about.

"New universe" is not hyperbole. Scientists guess the planet could harbor more than a million yet-to-be-discovered viruses. Bacteria are even trickier, and so we probably know about even fewer of them.

Perhaps scariest are those that live within us, peacefully for now. More than 99 percent of even those bacteria inside human bodies are presently unknown to science, which means we are operating in near-total ignorance about the effects climate change might have on the bugs in, for instance, our guts—about how many of the bacteria modern humans have come to rely on, like unseen factory workers, for everything from digesting our food to modulating our anxiety, could be rewired, diminished, or entirely killed off by an additional few degrees of heat.

Overwhelmingly, of course, the viruses and bacteria making

homes inside us are nonthreatening to humans—at present. Presumably, a difference of a degree or two in global temperature won't dramatically change the behavior of the majority of them—probably the vast majority, even the overwhelming majority. But consider the case of the saiga—the adorable, dwarflike antelope, native to central Asia. In May 2015, nearly two-thirds of the global population died in the span of just days—every single saiga in an area the size of Florida, the land suddenly dotted with hundreds of thousands of saiga carcasses and not one lone survivor. An event like this is called a "mega-death," this one so striking and cinematic that it gave rise, immediately, to a whole raft of conspiracy theories: aliens, radiation, dumped rocket fuel. But no toxins were found by researchers poking through the killing fields—in the animals themselves, in the soil, in the local plants. The culprit, it turned out, was a simple bacteria, *Pasteurella multocida*, which had lived inside the saiga's tonsils, without threatening its host in any way, for many, many generations. Suddenly it had proliferated, emigrated to the bloodstream, and from there to the animals' liver, kidneys, and spleen. Why? "The places where the saigas died in May 2015 were extremely warm and humid," Ed Yong wrote in *The Atlantic*. "In fact, humidity levels were the highest ever seen in the region since records began in 1948. The same pattern held for two earlier, and much smaller, die-offs from 1981 and 1988. When the temperature gets really hot, and the air gets really wet, saiga die. Climate is the trigger, *Pasteurella* is the bullet."

This is not to say we now understand what precisely about humidity weaponized *Pasteurella*, or how many of the other bacteria living inside mammals like us—the 1 percent we have identified, or perhaps more worryingly the 99 percent we house without any knowledge or understanding—might be similarly triggered by climate, friendly, symbiotic bugs with whom we've lived in some cases for millions of years, transformed suddenly into contagions already inside us. That remains a mystery. But ignorance is no comfort. Presumably climate change will introduce us to some of them.

Economic
Collapse

THE MURMURING MANTRA OF GLOBAL MARKETS—WHICH prevailed between the end of the Cold War and the onset of the Great Recession, promising something like their own eternal reign—is that economic growth will save us from anything and everything.

But in the aftermath of the 2008 crash, a number of historians and iconoclastic economists studying what they call "fossil capitalism" have started to suggest that the entire history of swift economic growth, which began somewhat suddenly in the eighteenth century, is not the result of innovation or the dynamics of free trade, but simply our discovery of fossil fuels and all their raw power—a onetime injection of that new "value" into a system that had previously been characterized by unending subsistence living. This is a minority view, among economists, and yet the précis version of the perspective is quite powerful. Before fossil fuels, nobody lived better than their parents or grandparents or ancestors from five hundred years before, except in the immediate aftermath of a great plague like the Black Death, which allowed the lucky survivors to gobble up the resources liberated by mass graves.

In the West especially, we tend to believe we've invented our way out of that endless zero-sum, scratch-and-claw resource scramble—both with particular innovations, like the steam engine and computer, and with the development of a dynamic capitalistic system to reward them. But scholars like Andreas Malm have a different perspective: we have been extracted from that muck only by a singular innovation, one engineered not by entrepreneurial human hands but in fact millions of years before the first ones ever dug at the earth—engineered by time and geologic weight, which many millennia ago pressed the fossils of Earth's earlier carbon-based life forms (plants, small animals) into petroleum, like lemon under a press. Oil is the patrimony of the planet's prehuman past: what stored energy the earth can produce when undisturbed for millennia. As soon as humans discovered that storehouse, they set about plundering it—so fast that, at various points over the last half century, oil forecasters have panicked about running out. In 1968, the labor historian Eric Hobsbawm wrote, "Whoever says Industrial Revolution, says cotton." Today, he would probably substitute "fossil fuel."

The timeline of growth is just about perfectly consistent with the burning of those fuels, though doctrinaire economists would argue there is much more to the equation of growth. Generations being as long as they are and historical memory as short, the West's several centuries of relatively reliable and expanding prosperity have endowed economic growth with the reassuring aura of permanence: we expect it, on some continents, at least, and rage against our leaders and elites when it does not come. But planetary history is very long, and human history, though a briefer interval, is long, too. And while the pace of technological change we call progress is today dizzying and may yet invent new ways of buffering us from the blows of climate change, it is also not hard to imagine those flush centuries, enjoyed by nations who colonized the rest of the planet to produce them, as an aberration. Earlier empires had boom years, too.

You do not have to believe that economic growth is a mirage produced by fossil fumes to worry that climate change is a threat to it—in fact, this proposition forms the cornerstone around which an entire edifice of academic literature has been built over the last decade. The most exciting research on the economics of warming has come from Solomon Hsiang and Marshall Burke and Edward Miguel, who are not historians of fossil capitalism but who offer some very bleak analysis of their own: in a country that's already relatively warm, every degree Celsius of warming reduces growth, on average, by about one percentage point (an enormous number, considering we count growth in the low single digits as "strong"). This is the sterling work in the field. Compared to the trajectory of economic growth with no climate change, their average projection is for a 23 percent loss in per capita earning globally by the end of this century.

Tracing the shape of the probability curve is even scarier. There is a 51 percent chance, this research suggests, that climate change will reduce global output by more than 20 percent by 2100, compared with a world without warming, and a 12 percent chance that it lowers per capita GDP by 50 percent or more by then, unless emissions decline. By comparison, the Great Depression dropped global GDP by about 15 percent, it is estimated—the numbers weren't so good back then. The more recent Great Recession lowered it by about 2 percent, in a onetime shock; Hsiang and his colleagues estimate a one-in-eight chance of an ongoing and irreversible effect by 2100 that is twenty-five times worse. In 2018, a team led by Thomas Stoerk suggested that these estimates could be dramatic underestimates.

The scale of that economic devastation is hard to comprehend. Even within the postindustrial nations of the wealthy West, where economic indicators such as the unemployment rate and GDP growth circulate as though they contain the whole meaning of life in them, figures like these are a little bit hard to fathom; we've become so used to economic stability and reliable growth that the entire spectrum of conceivability stretches from contractions of about 15 percent, effects we study still in histories of the Depression, to growth about half as fast—about 7 percent, which the world as a whole last achieved during the global boom of the early 1960s. These are exceptional onetime peaks and troughs, extending for no more than a few years, and most of the time we measure economic fluctuations in ticks of decimal points—2.9 this year, 2.7 that. What climate change proposes is an economic setback of an entirely different category.

The breakdown by country is perhaps even more alarming. There are places that benefit, in the north, where warmer temperatures can improve agriculture and economic productivity: Canada, Russia, Scandinavia, Greenland. But in the mid-latitudes, the countries that produce the bulk of the world's economic activity— the United States, China—lose nearly half of their potential output. The warming near the equator is worse, with losses throughout Africa, from Mexico to Brazil, and in India and Southeast Asia approaching 100 percent. India alone, one study proposed, would shoulder nearly a quarter of the economic suffering inflicted on the entire world by climate change. In 2018, the World Bank estimated that the current path of carbon emissions would sharply diminish the living conditions of 800 million living throughout South Asia. One hundred million, they say, will be dragged into extreme poverty by climate change just over the next decade. Perhaps "back into" is more appropriate: many of the most vulnerable are those populations that have just extracted themselves from deprivation and subsistence living, through developing-world growth powered by industrialization and fossil fuel.

And to help buffer or offset the impacts, we have no New Deal revival waiting around the corner, no Marshall Plan ready. The global halving of economic resources would be permanent, and, because permanent, we would soon not even know it as deprivation, only as a brutally cruel normal against which we might measure tiny burps of decimal-point growth as the breath of a new prosperity. We have gotten used to setbacks on our erratic march along the arc of economic history, but we know them as setbacks and expect elastic recoveries. What climate change has in store is not that kind of thing—not a Great Recession or a Great Depression but, in economic terms, a Great Dying.

How could that come to be? The answer is partly in the preceding chapters—natural disaster, flooding, public health crises. All of these are not just tragedies but expensive ones, and beginning already to accumulate at an unprecedented rate. There is the cost to agriculture: more than three million Americans work on more than two million farms; if yields decline by 40 percent, margins will decline, too, in many cases disappearing entirely, the small farms and cooperatives and even empires of agribusinesses slipping underwater (to use the oddly apposite accountant's metaphor) and drowning under debt all those who own and work those arid fields, many of them old enough to remember the same plains' age of plenty. And then there is the real flooding: 2.4 million American homes and businesses, representing more than $1 trillion in present-day value, will suffer chronic flooding by 2100, according to a 2018 study by the Union of Concerned Scientists. Fourteen percent of the real estate in Miami Beach could be flooded by just 2045. This is just within America, though it isn't only South Florida; in fact, over the next few decades, the real-estate impact will be almost $30 billion in New Jersey alone.

There is a direct heat cost to growth, as there is to health. Some of these effects we can see already—for instance, the warping of train tracks or the grounding of flights due to temperatures so high that they abolish the aerodynamics that allow planes to take off, which is now commonplace at heat-stricken airports like the one in Phoenix. (Every round-trip plane ticket from New York to London, keep in mind, costs the Arctic three more square meters of ice.) From Switzerland to Finland, heat waves have necessitated the closure of power plants when cooling liquids have become too hot to do their job. And in India, in 2012, 670 million lost power when the country's grid was overwhelmed by farmers irrigating their fields without the help of the monsoon season, which never arrived. In all but the shiniest projects in all but the wealthiest parts of the world, the planet's infrastructure was simply not built for climate change, which means the vulnerabilities are everywhere you look.

Other, less obvious effects are also visible—for instance, productivity. For the past few decades, economists have wondered why the computer revolution and the internet have not brought meaningful productivity gains to the industrialized world. Spreadsheets, database management software, email—these innovations alone would seem to promise huge gains in efficiency for any business or economy adopting them. But those gains simply haven't materialized; in fact, the economic period in which those innovations were introduced, along with literally thousands of similar computer-driven efficiencies, has been characterized, especially in the developed West, by wage and productivity stagnation and dampened economic growth. One speculative possibility: computers have made us more efficient and productive, but at the same time climate change has had the opposite effect, diminishing or wiping out entirely the impact of technology. How could this be? One theory is the negative cognitive effects of direct heat and air pollution, both of which are accumulating more research support by the day. And whether or not that theory explains the great stagnation of the last several decades, we do know that, globally, warmer temperatures do dampen worker productivity.

The claim seems both far-fetched and intuitive, since, on the one hand, you don't imagine a few ticks of temperature would turn entire economies into zombie markets, and since, on the other, you yourself have surely labored at work on a hot day with the air-conditioning out and understand how hard that can be. The bigger-picture perspective is harder to swallow, at least at first. It may sound like geographic determinism, but Hsiang, Burke, and Miguel have identified an optimal annual average temperature for economic productivity: 13 degrees Celsius, which just so happens to be the historical median for the United States and several other of the world's biggest economies. Today, the U.S. climate hovers around 13.4 degrees, which translates into less than 1 percent of GDP loss—though, like compound interest, the effects grow over time. Of course, as the country has warmed over the last decades, particular regions have seen their temperatures rise, some of them from suboptimal levels to something closer to an ideal setting, climate-wise. The greater San Francisco Bay Area, for instance, is sitting pretty right now, at exactly 13 degrees.

This is what it means to suggest that climate change is an enveloping crisis, one that touches every aspect of the way we live on the planet today. But the world's suffering will be distributed as unequally as its profits, with great divergences both between countries and within them. Already-hot countries like India and Pakistan will be hurt the most; within the United States, the costs will be shouldered largely in the South and Midwest, where some regions could lose up to 20 percent of county income.

Overall, though it will be hit hard by climate impacts, the United States is among the most well-positioned to endure—its wealth and geography are reasons that America has only begun to register effects of climate change that already plague warmer and poorer parts of the world. But in part because it has so much to lose, and in part because it so aggressively developed its very long coastlines, the U.S. is more vulnerable to climate impacts than any country in the world but India, and its economic illness won't be quarantined at the border. In a globalized world, there is what Zhengtao Zhang and others call an "economic ripple effect." They've also quantified it, and found that the impact grows along with warming. At one degree Celsius, with a decline in American GDP of 0.88 percent, global GDP would fall by 0.12 percent, the American losses cascading through the world system. At two degrees, the economic ripple effect triples, though here, too, the effects play out differently in different parts of the world; compared to the impact of American losses at one degree, at two degrees the economic ripple effect in China would be 4.5 times larger. The radiating shock waves issuing out from other countries are smaller because their economies are smaller, but the waves will be coming from nearly every country in the world, like radio signals beamed out from a whole global forest of towers, each transmitting economic suffering.

For better or for worse, in the countries of the wealthy West we have settled on economic growth as the single best metric, however imperfect, of the health of our societies. Of course, using that metric, climate change registers—with its wildfires and droughts and famines, it registers seismically. The costs are astronomical already, with single hurricanes now delivering damage in the hundreds of billions of dollars. Should the planet warm 3.7 degrees, one assessment suggests, climate change damages could total $551 trillion—nearly twice as much wealth as exists in the world today. We are on track for more warming still.

Over the last several decades, policy consensus has cautioned that the world would only tolerate responses to climate change if they were free—or, even better, if they could be presented as avenues of economic opportunity. That market logic was probably always shortsighted, but over the last several years, as the cost of adaptation in the form of green energy has fallen so dramatically, the equation has entirely flipped: we now know that it will be much, much more expensive to *not* act on climate than to take even the most aggressive action today. If you don't think of the price of a stock or government bond as an insurmountable barrier to the returns you'll receive, you probably shouldn't think of climate adaptation as expensive, either. In 2018, one paper calculated the global cost of a rapid energy transition, by 2030, to be negative $26 trillion—in other words, rebuilding the energy infrastructure of the world would make us all that much money, compared to a static system, in only a dozen years.

Every day we do not act, those costs accumulate, and the numbers quickly compound. Hsiang, Burke, and Miguel draw their 50 percent figure from the very high end of what's possible—truly a worst-case scenario for economic growth under the sign of climate change. But in 2018, Burke and several other colleagues published a major paper exploring the growth consequences of some scenarios closer to our present predicament. In it, they considered one plausible but still quite optimistic scenario, in which the world meets its Paris Agreement commitments, limiting warming to between 2.5 and 3 degrees. This is probably about the best-case warming scenario we might reasonably expect; globally, relative to a world with no additional warming, it would cut per-capita economic output by the end of the century, Burke and his colleagues estimate, by between 15 and 25 percent. Hitting four degrees of warming, which lies on the low end of the range of warming implied by our current emissions trajectory, would cut into it by 30 percent or more. This is a trough twice as deep as the deprivations that scarred our grandparents in the 1930s, and which helped produce a wave of fascism, authoritarianism, and genocide. But you can only really call it a trough when you climb out of it and look back from a new peak, relieved. There may not be any such relief or reprieve from climate deprivation, and though, as in any collapse, there will be those few who find ways to benefit, the experience of most may be more like that of miners buried permanently at the bottom of a shaft.

Climate Conflict

C LIMATOLOGISTS ARE VERY CAREFUL WHEN TALKING ABOUT
Syria. They want you to know that while climate change did
produce a drought that contributed to the country's civil war, it is
not exactly fair to say that the conflict is the result of warming;
next door, for instance, Lebanon suffered the same crop failures
and remained stable.

But wars are not caused by climate change only in the same way
that hurricanes are not caused by climate change, which is to say
they are made more likely, which is to say the distinction is semantic. If climate change makes conflict only 3 percent more likely in
a given country, that does not mean it is a trivial effect: there are
almost two hundred countries in the world, which multiplies the
likelihood, meaning that rise in temperature could yield three or
four or six more wars. Over the last decade, researchers have even
managed to quantify some of the nonobvious relationships between
temperature and violence: for every half degree of warming, they

say, societies will see between a 10 and 20 percent increase in the likelihood of armed conflict. In climate science, nothing is simple, but the arithmetic is harrowing: a planet four degrees warmer would have perhaps twice as many wars as we do today. And likely more.

As is the case with nearly every aspect of climate chaos, meeting the Paris goals will not save us from this bloodshed, in fact far from it; even an astonishing, improbable effort to limit warming to two degrees would still, by this math, result in at least 40 percent, and perhaps as much as 80 percent, more war. This, in other words, is our best-case scenario: at least half again as much conflict as we see today, when few watching the news each night would say we are enjoying an abundance of peace. Already, climate change has elevated Africa's risk of conflict by more than 10 percent; in that continent, by just 2030, projected temperatures are expected to cause 393,000 additional deaths in battle.

"BATTLE"—THE WORD FEELS LIKE A RELIC WHEN YOU COME across it. In the wealthy West, we've come to pretend that war is an anomalous feature of modern life, since it seems to have been retired as fully from our everyday experience as polio. But globally, there are nineteen ongoing armed conflicts hot enough to claim at least a thousand lives each year. Nine of them began more recently than 2010, and many more unfold at smaller scales of violence.

That all of these counts are expected to spike in the coming decades is one reason that, as nearly every climate scientist I've spoken to has pointed out, the U.S. military is obsessed with climate change, the Pentagon issuing regular climate threat assessments and planning for a new era of conflict governed by global warming. (This is still true in the Trump era, when lesser federal outfits like the Government Accountability Office deliver grim warnings about climate, too.) The drowning of American navy bases by sea-level rise is trouble enough, and the melting of the Arctic promises to open an entirely new theater of conflict, once nearly as foreign-seeming as the space race. (It also positions the country primarily against America's old rivals the Russians, now revived as adversaries.)

Given the right war-gaming cast of mind, it is also possible to see the aggressive Chinese construction activity in the South China Sea, where whole new artificial islands have been erected for military use, as a kind of dry run, so to speak, for life as a super-power in a flooded world. The strategic opportunity is clear, with so many of the existing footholds—like all those low-lying islands the United States once used to stepping-stone its own empire across the Pacific—expected to disappear by the end of the century, if not before. The Marshall Islands archipelago, for instance, seized by the U.S. during World War II, could be rendered uninhabitable by sea-level rise as soon as midcentury, the U.S. Geological Survey has warned; its islands will be underwater even if we meet the Paris goals. And what is taken down with them is quite scary. Beginning with the bombing at Bikini Atoll, these islands were ground zero for American atom bomb testing just after the war; the U.S. military has only ever "cleaned up" one island of radioactivity, which makes them the world's largest nuclear waste site.

But for the military, climate change is not just a matter of great-power rivalry executed across a transformed map. Even for those in the American military who expect the country's hegemony to endure indefinitely, climate change presents a problem, because being the world's policeman is quite a bit harder when the crime rate doubles. And it's not just Syria where climate has contributed to conflict. Some speculate that the elevated level of strife across the Middle East over the past generation reflects the pressures of global warming—a hypothesis all the more cruel considering that warming began to accelerate when the industrialized world extracted and then burned the region's oil. From Boko Haram to ISIS to the Taliban and militant Islamic groups in Pakistan, drought and crop failure have been linked to radicalization, and the effect may be especially pronounced amid ethnic strife: from 1980 to 2010, a 2016 study found, 23 percent of conflict in the world's ethnically diverse countries began in months stamped by weather disaster. According to one assessment, thirty-two countries—from Haiti to the Philippines and India to Cambodia, each heavily dependent on farming and agriculture—face "extreme risk" of conflict and civil unrest from climate disruptions over the next thirty years.

What accounts for the relationship between climate and conflict? Some of it comes down to agriculture and economics: when yields drop and productivity falls, societies can falter, and when droughts and heat waves hit, the shocks can be felt even more deeply, electrifying political fault lines and producing or exposing others no one knew to worry over. A lot has to do with the forced migration that can result from those shocks, and with the political and social instability that migration often produces; when things go south, those who are able tend to flee, not always to places ready to welcome them—in fact, recent history shows, often quite the opposite. And today migration is already at a record high, with almost seventy million displaced people wandering the planet right now. That is the outbound impact; but the local one is often more profound. Those who remain in a region ravaged by extreme weather often find themselves navigating an entirely new social and political structure, if one endures at all. And it is not just weak states that can fall at the hands of climate pressures; in recent years, scholars have compiled a long list of empires buckled, at least in part, by climate effects and events: Egypt, Akkadia, Rome.

This complex calculus is what makes researchers reluctant to assign blame for conflict neatly, but complexity is how warming articulates its brutality. Like the cost to growth, war is not a discrete impact of global temperature rise but something more like an all-encompassing aggregation of climate change's worst tremors and cascades. The Center for Climate and Security, a state-focused think tank, organizes the threats from climate change into six categories: "Catch-22 states," in which governments have responded to local climate challenges—to agriculture, for example—by turning toward a global marketplace that is now more than ever vulnerable to climate shocks; "brittle states," stable on the surface—but only by a run of good climate luck; "fragile states," such as Sudan, Yemen, and Bangladesh, where climate impacts have already eaten into trust in state authority, or worse; "disputed zones among states," like the South China Sea or Arctic; "disappearing states," which they mean literally, as in the case of the Maldives; and "non-state actors," like ISIS, which can seize local resources, such as freshwater, as a way of applying leverage against the nominal state authority or the local population. In each case, climate is not the sole cause but the spark igniting a complex bundle of social kindling.

This complexity may also be one reason we cannot see the threat of escalating war very clearly, choosing to regard conflict as something determined primarily by politics and economics when all three are in fact governed, like everything else, by the conditions established by our rapidly changing climate. Over the last decade or so, the linguist Steven Pinker has made a second career out of suggesting that, in the West especially, we are unable to appreciate human progress—are in fact blind to all of the massive and rapid improvements the world has witnessed in less violence and war and poverty, reduced infant mortality, and enhanced life expectancy. It's true, we are. When you look at the charts, the trajectory of that progress seems inarguable: so many fewer violent deaths, so much less extreme deprivation, a global middle class expanding by the hundreds of millions. But again, that story is about the wealth brought by industrialization and the transformations of societies by newfound wealth powered by fossil fuel. It is a story written largely by China and, to a lesser extent, the rest of the developing world, which has developed by industrializing. And the cost of much of that progress, the balance come due for all the industrialization that made middle-class-ness possible for the billions of people in the global south, is climate change—which we are, ironically, far too sanguine about, Pinker included. Worse still, the warming unleashed by all our progress heralds a return to violence.

Even when it comes to war, historical memory has a sadistically short half-life, horrors and their causes gauzily evanescing into familiar folklore in less than the span of a single generation. But most wars throughout history, it is important to remember, have been conflicts over resources, often ignited by resource scarcity, which is what an earth densely populated and denuded by climate change will yield. Those wars don't tend to increase those resources; most of the time, they incinerate them.

THE FOLKLORE OF STATE CONFLICT CASTS A LONG SHADOW— the patchwork quilt of nations tugged apart into a ghastly, mutually damaging disarray. Climate tugs at the individual threads of conflict, too: personal irritability, interpersonal conflict, domestic violence.

Heat frays everything. It increases violent crime rates, swearing on social media, and the likelihood that a major-league pitcher, coming to the mound after his teammate has been hit by a pitch, will hit an opposing batter in retaliation. The hotter it gets, the longer drivers will honk their horns in frustration; and even in simulations, police officers are more likely to fire on intruders when the exercises are conducted in hotter weather. By 2099, one speculative paper tabulated, climate change in the United States would bring about an additional 22,000 murders, 180,000 rapes, 3.5 million assaults, and 3.76 million robberies, burglaries, and acts of larceny. The statistics of the past are more inarguable, and even the arrival of air-conditioning in the developed world in the middle of the last century did little to solve the problem of the summer crime wave.

It's not just temperature effects. In 2018, a team of researchers examining an enormous data set of more than 9,000 American cities found that air pollution levels positively predicted incidents of every single crime category they looked at—from car theft and burglary and larceny up to assault, rape, and murder. And then there are the ways that climate impacts can cascade into violence more circuitously. Between 2008 and 2010, Guatemala was hit by Tropical Storm Arthur, Hurricane Dolly, Tropical Storm Agatha, and Tropical Storm Hermine—this a country that was already one of the ten most affected by extreme weather and reeling in the same years from the eruption of a local volcano and a regional earthquake. All told, almost three million were left "food insecure," and at least 400,000 needed humanitarian assistance; from the 2010 disasters alone, the country sustained damages totaling more than a billion dollars, or roughly a quarter of the national budget, devastating its roads and supply chains. In 2011, it was hit by Tropical Storm 12E, and, in the wake of the disasters, farmers turned to growing poppies; organized crime, already an enormous problem, exploded—which should perhaps not surprise us, given that recent research has shown that the Sicilian mafia was produced by drought. Today, Guatemala has the fifth-highest homicide rate in the world; according to UNICEF, it is the second most dangerous country in the world for children. Historically, the country's cash crops have been coffee and sugarcane; in the coming decades, climate change could make both of them ungrowable.

"Systems"

W HAT I CALL CASCADES, CLIMATE SCIENTISTS CALL "SYS-
tems crises." These crises are what the American military
means when it names climate change a "threat multiplier." The
multiplication, when it falls short of conflict, produces migration—
that is, climate refugees. Since 2008, by one count, it has already
produced 22 million of them.

In the West, we often think of refugees as a failed-state
problem—that is, a problem that the broken and impoverished
parts of the world inflict on relatively more stable, and wealthier,
societies. But Hurricane Harvey produced at least 60,000 climate
migrants in Texas, and Hurricane Irma forced the evacuation
of nearly 7 million. As with so much else, it will only get worse
from here. By 2100, sea-level rise alone could displace 13 million
Americans—a few percent of the country's total population. Many
of those sea-level refugees will come from the country's southeast—
chiefly Florida, where 2.5 million are expected to be flooded out

of greater Miami; and Louisiana, where the New Orleans area is predicted to lose half a million.

As an unusually wealthy country, the United States is, for now, unusually suited to withstand such disruptions—one can almost imagine, over the course of a century, tens of millions of resettled Americans adapting to a ravaged coastline and a new geography for the country. Almost. But warming is not just a matter of sea level, and its horrors will not hit nations like the United States first. In fact, the impacts will be greatest in the world's least developed, most impoverished, and therefore least resilient nations—almost literally a story of the world's rich drowning the world's poor with their waste. The first country to industrialize and produce greenhouse gas on a grand scale, the United Kingdom, is expected to suffer least from climate change. The world's slowest-developing countries, producing the least emissions, will be among those hardest hit; the climate system of the Democratic Republic of Congo, one of the world's poorest countries, is scheduled to be especially profoundly perturbed.

The Congo is mostly landlocked, and mountainous, but in the next generation of warming those features will not be protections. Wealth will be a buffer for some countries, but not a safeguard, as Australia is learning already: by far the richest of all the countries staring down the most intense, most immediate warming barrages, it is an early test case of how the world's affluent societies will bend, or buckle, or rebuild under the pressure of temperature changes likely to hit the rest of the well-off world only later this century. The country was founded on genocidal indifference to the native landscape and those who inhabited it, and its modern ambitions have always been precarious: Australia is today a society of expansive abundance, jerry-rigged onto a very harsh and ecologically unforgiving land. In 2011, a single heat wave there produced significant tree dieback and coral bleaching, the death of plant life, crashes in local bird populations and dramatic spikes in the number of certain insects, and transformations of ecosystems both marine and terrestrial. When the country enacted a carbon tax, its emissions fell; when, under political pressure, the tax was repealed, they rose again. In 2018, the country's parliament declared global warming a "current and existential national security risk." A few months later, its climate-conscious prime minister was forced to resign, for the shame of attempting to honor the Paris accords.

The wheels of all communities are greased by abundance; baked by deprivation, they stall and crack. The paths are familiar ones, even to those who have only ever known affluence, their lives creamily frictionless but stimulated by entertainments tracing the arc of social decline: market breakdowns, price gouging, the hoarding of goods and services by the well-off and well-armed, the retreat of law enforcement into self-enrichment, and the disappearance of any expectation of justice making survival suddenly a matter of entrepreneurial skill.

More than 140 million people in just three regions of the world will be made climate migrants by 2050, the World Bank projected in a 2018 study, assuming current warming and emissions trends: 86 million in sub-Saharan Africa, 40 million in South Asia, and 17 million in Latin America. The most commonly cited estimate from the United Nations' International Organization for Migration suggests numbers a bit higher—200 million, total, by 2050. These figures are quite high—higher than most non-advocates credit. But according to the U.N. IOM, climate change may unleash as many as a billion migrants on the world by 2050. One billion—that is about as many people as live today in North and South America combined. Imagine the two continents suddenly drowned in the sea, the whole New World submerged, and everyone left bobbing at the surface now fighting for a foothold, somewhere, anywhere, and, if someone else is scrambling for the same dry spot, scrambling to get there first.

———

THE SYSTEM IN CRISIS IS NOT ALWAYS "SOCIETY"; THE SYSTEM can also be the body. Historically, in the United States, more than two-thirds of outbreaks of waterborne disease—illnesses smuggled into humans through algae and bacteria that can produce gastrointestinal problems—were preceded by unusually intense rainfall, disrupting local water supplies. The concentration of salmonella in streams, for instance, increases significantly after heavy rainfall, and the country's most dramatic outbreak of waterborne disease came in 1993, when more than 400,000 in Milwaukee fell ill from cryptosporidium immediately after a storm.

Sudden rainfall shocks—both deluges and their opposite, droughts—can devastate agricultural communities economically, but also produce what scientists call, with understatement, "nutritional deficiencies" in fetuses and infants; in Vietnam, those who passed through that crucible early on, and survived, were shown to start school later in life, do worse when they got there, and grow less tall than their peers. In India, the same cycle-of-poverty pattern holds. The lifelong impacts of chronic malnutrition are more troubling still for being permanent: diminishing cognitive ability, flattened adult wages, increased morbidity. In Ecuador, climate damage has been seen even in middle-class children, who bear the mark of rainfall shocks and extreme temperatures on their wages twenty to sixty years after the fact. The effects begin in the womb, and they are universal, with measurable declines in lifetime earnings for every day over ninety degrees during a baby's nine months in utero. The impacts accumulate later in life, too. An enormous study in Taiwan found that, for every single unit of additional air pollution, the relative risk of Alzheimer's doubled. Similar patterns have been observed from Ontario to Mexico City.

As conditions of environmental degradation become more universal, it may, perversely, require more imagination to consider their costs. When the deprived are no longer outlier communities but instead whole regions, whole countries, conditions that once may have seemed inhumane now appear, to a future generation who knows no better, simply "normal." In the past, we have looked in horror at the stunted growth of national populations who passed through famines both natural (Sudan, Somalia) and man-made (Yemen, North Korea). In the future, climate change may stunt us all, in one way or another, with no control group entirely spared.

You might expect these premonitions to settle like sediments into family planning. And indeed, among the young and well-off in Europe and the United States, for whom reproductive choices are often freighted with political meaning, they have. Among this outwardly conscientious cohort, there is much worry about bringing new children into a degraded world, full of suffering, and about "contributing" to the problem by crowding the climate stage with more players, each a little consumption machine. "Want to fight climate change?" *The Guardian* asked in 2017. "Have fewer children." That year and the next, the paper published several variations on the theme, as did many other publications delivered to the lifestyle class, including *The New York Times:* "Add this to the list of decisions affected by climate change: Should I have children?"

The effect on the personal decisions of the consumer class is perhaps a narrow way of thinking about global warming, though it demonstrates a strain of strange ascetic pride among the well-to-do. ("The egoism of child-bearing is like the egoism of colonizing a country," the novelist Sheila Heti writes, in a representative passage from *Motherhood,* her meditation on the meaning of parenthood, which she chose to avoid.) But of course further degradation isn't inescapable; it is optional. Each new baby arrives in a brand-new world, contemplating a whole horizon of possibilities. The perspective is not naive. We live in that world with them—helping make it for them, and with them, and for ourselves. The next decades are not yet determined. A new timer begins with every birth, measuring how much more damage will be done to the planet and the life this child will live on it. The horizons are just as open to us, however foreclosed and foreordained they may seem. But we close them off when we say anything about the future being inevitable. What may sound like stoic wisdom is often an alibi for indifference.

IN A WORLD OF SUFFERING, THE SELF-INTERESTED MIND CRAVES
compartmentalization, and one of the most interesting frontiers of
emerging climate science traces the imprint left on our psychological well-being by the force of global warming, which can over-
whelm whatever methods we devise to cope—that is, the mental
health effects of a world on fire. Perhaps the most predictable vector is trauma: between a quarter and a half of all those exposed to
extreme weather events will experience them as an ongoing negative shock to their mental health. In England, flooding was found
to quadruple levels of psychological distress, even among those in
an inundated community but not personally affected by the flooding. In the aftermath of Hurricane Katrina, 62 percent of evacuees
exceeded the diagnostic threshold for acute stress disorder; in the
region as a whole, nearly a third had PTSD. Wildfires, curiously,
yielded a lower incidence—just 24 percent of evacuees in the aftermath of one series of California blazes. But a third of those who
lived through fire were diagnosed, in its aftermath, with depression.

Even those watching the effects from the sidelines suffer from
climate trauma. "I don't know of a single scientist that's not having
an emotional reaction to what is being lost," Camille Parmesan,
who shared the 2007 Nobel Peace Prize with Al Gore, has said.
Grist has called the phenomenon "climate depression," *Scientific
American* "environmental grief." And while it may seem intuitive
that those contemplating the end of the world find themselves de-
spairing, especially when their calls of alarm have gone almost entirely unheeded, it is also a harrowing forecast of what is in store
for the rest of the world, as the devastation of climate change slowly
reveals itself. In the sense of psychological distress, which so many
of them endure, climate scientists are the canaries in our coal mine.
This may be why so many of them seem concerned with the risks
of crying wolf about warming: they've learned enough about public
apathy to worry themselves into knots about just when, and precisely how, to raise the alarm.

In certain places, that alarm has been raised for them. Those studying the phenomenon are only suffering secondhand—which is a sign of just how intense the firsthand impact has been. Unsurprisingly, climate trauma is especially harsh in the young—in this, our folk wisdom about the impressionable minds of children is reliable. Thirty-two weeks after Hurricane Andrew hit Florida in 1992, killing forty, more than half of children surveyed had moderate PTSD and more than a third had a severe form; in the high-impact areas, 70 percent of children scored in the moderate-to-severe range fully twenty-one months after the Category 5 storm. By dismal contrast, soldiers returning from war are estimated to suffer from PTSD at a rate between 11 and 31 percent.

One especially detailed study examined the mental health fallout from Hurricane Mitch, a Category 5 storm and the second-deadliest Atlantic hurricane on record, which struck Central America in 1998, leaving 11,000 dead. In Posoltega, the most hard-hit region of Nicaragua, children had a 27 percent chance of having been seriously injured, a 31 percent chance of having lost a family member, and a 63 percent chance of their home having been damaged or destroyed. You can imagine the aftereffects. Ninety percent of adolescents in the area were left with PTSD, with the average adolescent boy registering at the high end of the range of "severe" PTSD, and the average teenage girl registering over the threshold of "very severe." Six months after the storm, four out of every five teenage survivors from Posoltega suffered from depression; more than half, the study found, compulsively nursed what the authors called, a bit euphemistically, "vengeful thoughts."

And then there are the more surprising mental health costs. Climate affects both the onset and the severity of depression, *The Lancet* has found. Rising temperature and humidity are married, in the data, to emergency-room visits for mental health issues. When it's hotter out, psychiatric hospitals see spikes in proper inpatient admissions, as well. Schizophrenics, especially, are admitted at much higher rates when the temperatures are higher, and, inside those hospitals, ward temperature significantly increases symptom severity in schizophrenic patients. Heat waves bring waves of other things, too: mood disorders, anxiety disorders, dementia.

Heat produces violence and conflict between people, we know, and so it should probably not surprise us that it also generates a spike in violence against oneself. Each increase of a single degree Celsius in monthly temperature is associated with almost a percentage point rise of the suicide rate in the United States, and more than two percentage points in Mexico; an unmitigated emissions scenario could produce 40,000 additional suicides in these countries by 2050. One startling paper by Tamma Carleton has suggested that global warming is already responsible for 59,000 suicides, many of them farmers, in India—where one-fifth of all the world's suicides now occur, and where suicide rates have doubled since just 1980. When temperatures are already high, she found, a rise of just one additional degree, on a single day, will produce seventy additional corpses, each dead by the farmer's own hand.

IF YOU HAVE MADE IT THIS FAR, YOU ARE A BRAVE READER. ANY one of these twelve chapters contains, by rights, enough horror to induce a panic attack in even the most optimistic of those considering it. But you are not merely considering it; you are about to embark on living it. In many cases, in many places, we already are.

In fact, what is perhaps most remarkable about all of the research summarized to this point—concerning not only refugees, health, and mental health, but also conflict and food supply and sea level and all of the other elements of climate disarray—is that it is research emerging from the world we know today. That is, a world just one degree warmer; a world not yet deformed and defaced beyond recognition; a world bound largely by conventions devised in an age of climate stability, now barreling headlong into an age of something more like climate chaos, a world we are only beginning to perceive.

Some climate research is speculative, of course, projecting our best insights into physical processes and human dynamics onto planetary conditions no human being of any age or era has ever experienced. Some of these predictions will surely be falsified; that is how science proceeds. But all of our science arises from precedent, and the next era for climate change has none. The twelve elements of climate chaos are, as Donald Rumsfeld once put it in his incongruously useful phrasing, the "known knowns." This is the least concerning category; there are two more.

These sketches may feel exhaustive, at times even overwhelming. But they are merely sketches, to be filled in and fleshed out over the coming decades—if the previous decades are any guide, more often by bleaker science than by reassuring findings. For all our earned confidence in our knowledge of global warming—that it is real, that it is anthropogenic, that it is driving sea-level rise and Arctic melt and the rest—we still know only so much. Twenty years ago, there was no meaningful research on the relationship between climate change and economic growth; ten years ago, not much about climate and conflict. Fifty years ago, there was hardly any research about climate change whatsoever.

The pace of that scholarship is exhilarating, but it also counsels humility; there remains so much we do not know about the way global warming affects the way we live today. Now picture how much we'll know fifty years from now—and how much more gruesome our self-immolation will likely seem, even if we avoid its worst outcomes. Will warming trigger rapid feedback loops powered by the release of Arctic methane, or by the dramatic slowdown of the ocean's circulation system? It's impossible to say for sure. Will we protect ourselves by dispersing sulfur into our own now-red atmosphere, subjecting the entire planet to the uncertain health effects of those particles, or by erecting carbon-sucking plantations the size of continents? It's difficult to predict. These, then, are among the "known unknowns." And that oracle Rumsfeld furnished us with one conceptual category scarier still.

Which all means that these twelve threats described in these twelve chapters yield a portrait of the future only as best as it can be painted in the present. What actually lies ahead may prove even grimmer, though the reverse, of course, is also possible. The map of our new world will be drawn in part by natural processes that remain mysterious, but more definitively by human hands. At what point will the climate crisis grow undeniable, un-compartmentalizable? How much damage will have already been selfishly done? How quickly will we act to save ourselves and preserve as much of the way of life we know today as possible? For the sake of clarity, I've treated each of the threats from climate change—sea-level rise, food scarcity, economic stagnation—as discrete threats, which they are not. Some may prove offsetting, some mutually reinforcing, and others merely adjacent. But together they form a latticework of climate crisis, beneath which at least some humans, and probably many billions, will live. How?

III

The Climate
Kaleidoscope

Storytelling

I T SHOULD BE NO GREAT PRIZE TO BE RIGHT ABOUT THE END
of the world. But humans have told those stories incessantly,
across millennia, the lessons shifting with each imagined Armageddon. You'd think that a culture woven through with intimations
of apocalypse would know how to receive news of environmental
alarm. But instead we have responded to scientists channeling the
planet's cries for mercy as though they were simply crying wolf.
Today, the movies may be millenarian, but when it comes to contemplating real-world warming dangers, we suffer from an incredible failure of imagination. This is climate's kaleidoscope: we can
be mesmerized by the threat directly in front of us without ever
perceiving it clearly.

On-screen, climate devastation is everywhere you look, and
yet nowhere in focus, as though we are displacing our anxieties
about global warming by restaging them in theaters of our own
design and control—perhaps out of hope that the end of days

remains "fantasy." *Game of Thrones* opens with an unmistakable climate prophecy, but warns "winter is coming"; the premise of *Interstellar* is an environmental scourge, but the scourge is a crop blight. *Children of Men* depicts civilization in semi-collapse, but collapsed by a fertility menace. *Mad Max: Fury Road* unfurls like a global-warming panorama, a scrolling saga of a world made desert, but its political crisis comes, in fact, from an oil shortage. The protagonist of *The Last Man on Earth* is made that way by a sweeping virus, the family of *A Quiet Place* is hushed by giant insect predators lurking in the wilderness, and the central cataclysm of the "Apocalypse" season of *American Horror Story* is a throwback—a nuclear winter. In the many zombie apocalypses of this era of ecological anxiety, the zombies are invariably rendered as an alien force, not an endemic one. That is, not as us.

What does it mean to be entertained by a fictional apocalypse as we stare down the possibility of a real one? One job of pop culture is always to serve stories that distract even as they appear to engage—to deliver sublimation and diversion. In a time of cascading climate change, Hollywood is also trying to make sense of our changing relationship to nature, which we have long regarded from at least an arm's length—but which, amid this change, has returned as a chaotic force we nevertheless understand, on some level, as our fault. The adjudication of that guilt is another thing entertainment can do, when law and public policy fail, though our culture, like our politics, specializes in assigning the blame to others—in projecting rather than accepting guilt. A form of emotional prophylaxis is also at work: in fictional stories of climate catastrophe we may also be looking for catharsis, and collectively trying to persuade ourselves we might survive it.

Already, with the world just one degree warmer, wildfires and heat waves and hurricanes have inundated the news, and promise to cascade shortly through our stories and inner lives, making what may seem today a culture suffused with intuitions of doom look like a comparatively naive season. End-of-the-world nightmares will blossom, including in children's bedrooms, where siblings once whispered worries over the fact of death or the meaning of god-lessness or the possibility of protracted nuclear war; among their parents, climate trauma will take its place in the pop-psychological vernacular, if often as a scapegoat for more personal frustrations and anxieties. What will happen at two degrees, or three? Presumably, as climate change colonizes and darkens our lives and our world, it will do the same for our nonfiction, so much so that climate change may come to be regarded, at least by some, as the only truly serious subject.

In fictional narratives, in pop entertainments, and in what was once praised as "high" culture, a different, weirder course suggests itself. At first, perhaps a revival of the antiquated genre known as "Dying Earth"—initiated in English by Lord Byron with his poem "Darkness," written after a volcano eruption in the East Indies gave the Northern Hemisphere "The Year Without a Summer." That environmental alarm was echoed in similar fiction of the Victorian era, including H. G. Wells's *The Time Machine,* which depicted a distant future in which most humans were enslaved troglodytes, laboring underground for the benefit of a pampered and very small aboveground elite; in an even further future, almost all life on Earth had died. Our new version might include epic lamentations, a flourishing of what's been called already "climate existentialism." One scientist recently described to me the book she was working on as *"Between the World and Me* meets *The Road."*

But the scope of the world's transformation may just as quickly eliminate the genre—indeed eliminate any effort to narrativize warming, which could grow too large and too obvious even for Hollywood. You can tell stories "about" climate change while it still seems a marginal feature of human life, or an overwhelming feature of lives marginal to your own. But at three degrees of warming, or four, hardly anyone will be able to feel insulated from its impacts—or want to watch it on-screen as they watch it out their windows. And so as climate change expands across the horizon—as it begins to seem inescapable, total—it may cease to be a story and become, instead, an all-encompassing setting. No longer a narrative, it would recede into what literary theorists call metanarrative, succeeding those—like religious truth or faith in progress—that have governed the culture of earlier eras. This would be a world in which there isn't much appetite anymore for epic dramas about oil and greed, but where even romantic comedies would be staged under the sign of warming, as surely as screwball comedies were extruded by the anxieties of the Great Depression. Science fiction would be seen as even more prophetic, but the books that most eerily predicted the crisis will go unread, much like *The Jungle* or even *Sister Carrie* today: Why read about the world you can see plainly out your own window? At the moment, stories illustrating global warming can still offer an escapist pleasure, even if that pleasure often comes in the form of horror. But when we can no longer pretend that climate suffering is distant—in time or in place—we will stop pretending about it and start pretending within it.

IN HIS BOOK-LENGTH ESSAY *THE GREAT DERANGEMENT*, THE Indian novelist Amitav Ghosh wonders why global warming and natural disaster haven't yet become preoccupations of contemporary fiction, why we don't seem able to adequately imagine real-world climate catastrophe, why fiction hasn't yet made the dangers of warming sufficiently "real" to us, and why we haven't had a spate of novels in the genre he basically imagines into half existence and names "the environmental uncanny."

Others call it "cli-fi": genre fiction sounding environmental alarm, didactic adventure stories, often preachy in their politics. Ghosh has something else in mind: the great climate novel. "Consider, for example, the stories that congeal around questions like, 'Where were you when the Berlin Wall fell?' or 'Where were you on 9/11?'" he writes. "Will it ever be possible to ask, in the same vein, 'Where were you at 400 ppm?' or 'Where were you when the Larsen B ice shelf broke up?'"

His answer: Probably not, because the dilemmas and dramas

of climate change are simply incompatible with the kinds of stories we tell ourselves about ourselves, especially in conventional novels, which tend to end with uplift and hope and to emphasize the journey of an individual conscience rather than the miasma of social fate. This is a narrow definition of the novel, but almost everything about our broader narrative culture suggests that climate change is a major mismatch of a subject for all the tools we have at hand. Ghosh's question applies even to comic-book movies that might theoretically illustrate global warming: Who would the heroes be? And what would they be doing? The puzzle probably helps explain why so many pop entertainments that do try to tackle climate change, from *The Day After Tomorrow* on, are so corny and pedantic: collective action is, dramatically, a snore.

The problem is even more acute in gaming, which is poised to join or even supplant novels and movies and television, and which is built, as a narrative genre, even more obsessively around the imperatives of the protagonist—i.e., you. It also promises at least a simulation of agency. That could grow more comforting in the coming years, assuming we continue to proceed, zombie-like ourselves, down a path to ruin. Already, the world's most popular game, *Fortnite*, invites players into a competition for scarce resources during an extreme weather event—as though you yourself might conquer and totally resolve the issue.

There is also, beyond the hero problem, a villain problem. Literary fiction may not accommodate epic stories of the kind for which climate change fashions a natural setting, but, in the genre fiction and blockbuster movie space at least, we have a number of models at hand, from superhero sagas to alien-invasion narratives. Stories don't get more elemental and familiar than those that used to be described as "man against nature." But in *Moby-Dick* or *The Old Man and the Sea* or many lesser examples, nature was typically a metaphor, encasing a theological or metaphysical force. That was because nature remained mysterious, inexplicable. Climate change has changed that, too. We know the meaning of extreme weather and natural disaster, now, though they still arrive with a kind of prophetic majesty: the meaning is that there is more to come, and that it is our doing. You wouldn't have to do much in rewrites to *Independence Day* to reboot it as cli-fi. But, in the place of aliens, who would its heroes be fighting against? Ourselves?

Villainy was easier to grasp in stories depicting the prospect of nuclear Armageddon, the intuitive analogy to climate change, which ruled American culture for a generation. That was the whole cartoon of *Dr. Strangelove*—that the fate of the world sat in the hands of a few insane people; if it all blew up, we'd know exactly who to blame. That moral clarity was not Stanley Kubrick's, or a projection of his nihilism, but something like the opposite: conventional wisdom about geopolitics in the then-adolescent nuclear age. The same logic of responsibility appeared in *Thirteen Days,* Robert Kennedy's memoir of the Cuban missile crisis, which endured in part because it comported so neatly with the lived experience of its average reader through those weeks in 1962: watching the prospect of global annihilation wax and wane in a protracted game of telephone being played by two men and their relatively small staffs.

The moral responsibility of climate change is much murkier. Global warming isn't something that might happen, should several people make some profoundly shortsighted calculations; it is something that is already happening, everywhere, and without anything like direct supervisors. Nuclear Armageddon, in theory, has a few dozen authors; climate catastrophe has billions of them, with responsibility distended over time and extended across much of the planet. This is not to say it is distributed evenly: though climate change will be given its ultimate dimensions by the course of industrialization in the developing world, at present the world's wealthy possess the lion's share of guilt—the richest 10 percent producing half of all emissions. This distribution tracks closely with global income inequality, which is one reason that many on the Left point to the all-encompassing system, saying that industrial capitalism is to blame. It is. But saying so does not name an antagonist; it names a toxic investment vehicle with most of the world as stakeholders, many of whom eagerly bought in. And who in fact quite enjoy their present way of life. That includes, almost certainly, you and me and everyone else buying escapism with our Netflix subscription. Meanwhile, it simply isn't the case that the socialist countries of the world are behaving more responsibly, with carbon, nor that they have in the past.

Complicity does not make for good drama. Modern morality plays need antagonists, and the desire gets stronger when apportioning blame becomes a political necessity, which it surely will. This is a problem for stories both fictional and non-, each kind drawing logic and energy from the other. The natural villains are the oil companies—and in fact a recent survey of movies depicting climate apocalypse found the plurality were actually about corporate greed. But the impulse to assign them full responsibility is complicated by the fact that transportation and industry make up less than 40 percent of global emissions. The companies' disinformation-and-denial campaigns are probably a stronger case for villainy—a more grotesque performance of corporate evilness is hardly imaginable, and, a generation from now, oil-backed denial will likely be seen as among the most heinous conspiracies against human health and well-being as have been perpetrated in the modern world. But evilness is not the same as responsibility, and climate denialism has captured just one political party in one country in the world—a country with only two of the world's ten biggest oil companies. American inaction surely slowed global progress on climate in a time when the world had only one superpower. But there is simply nothing like climate denialism beyond the U.S. border, which encloses the production of only 15 percent of the world's emissions. To believe the fault for global warming lies exclusively with the Republican Party or its fossil fuel backers is a form of American narcissism.

That narcissism, I suspect, will be broken by climate change. In the rest of the world, where action on carbon is just as slow and resistance to real policy changes just as strong, denial is simply not a problem. The corporate influence of fossil fuel is present, of course, but so are inertia and the allure of near-term gains and the preferences of the world's workers and consumers, who fall somewhere on a long spectrum of culpability stretching from knowing selfishness through true ignorance and reflexive, if naive, complacency. How do you narrativize that?

BEYOND THE MATTER OF VILLAINY IS THE STORY OF NATURE and our relationship to it. That story seemed for a very long time to be contained within the simple logic of parables and allegories. Climate change promises to transform everything we thought we knew about nature, including the moral infrastructure of those tales. We still tell them at nearly every age, from the animated movies toddlers watch before they learn the alphabet, to fairy tales lifted from earlier eras, to disaster movies and magazine features about the fate of endangered species, and segments on the nightly news about extreme weather, which rarely mention warming.

Parables are a teaching tool and work like glass dioramas in natural history museums: you pass by, you look, you believe that what is contained in the taxidermy scene has something to teach you—but only by the logic of metaphor, because you are not a stuffed animal and do not live in the scene but beyond it, outside it, observing rather than participating. The logic is twisted by global warming, because it collapses the perceived distance between humans and nature—between you and the diorama. One message of climate change is: you do not live outside the scene but within it, subject to all the same horrors you can see afflicting the lives of animals. In fact, warming is already hitting humans so hard that we shouldn't need to look elsewhere, to endangered species and imperiled ecosystems, in order to trace the progress of climate's horrible offensive. But we do, saddened by stranded polar bears and stories of struggling coral reefs. When it comes to climate parables, we tend to like best the ones starring animals, who are mute when we do not project our voices onto them, and who are dying, at our own hands—half of them extinct, E. O. Wilson estimates, by 2100.

Even as we face crippling impacts from climate on human life, we still look to those animals, in part because what John Ruskin memorably called the "pathetic fallacy" still holds: it can be curiously easier to empathize with them, perhaps because we would rather not reckon with our own responsibility, but instead simply feel their pain, at least briefly. In the face of a storm kicked up by humans, and which we continue to kick up each day, we seem most comfortable adopting a learned posture of powerlessness.

PLASTIC PANIC IS ANOTHER EXEMPLARY CLIMATE PARABLE, IN that it is also a climate red herring. The panic arises from the admirable desire to leave a smaller imprint on the planet, and a natural horror that the environment is so polluted by detritus passing through our air, our food, our flesh—in this way, it draws on a very modern obsession with hygiene and lightness as a form of consumer grace (an obsession familiar from recycling). But while plastics have a carbon footprint, plastic pollution is simply not a global warming problem—and yet it has slid into the center of our vision, at least briefly, the ban on plastic straws occluding, if only for a moment, the much bigger and much broader climate threat.

Another such parable is bee death. Beginning in 2006, curious readers were introduced to a new environmental fable, as American honeybee colonies began to suffer an almost annual mass die-off: 36 percent dead one year; 29 the next; 46 the next; 34 the next. As anyone with a calculator might have figured, the numbers didn't add up: if that many bee colonies collapsed each year, the total number would be very rapidly approaching zero, not steadily increasing, which it was. This was because beekeepers, who were mostly not adorable amateurs but industrial-scale livestock managers trucking their bees across the country in an endless loop of pollination for hire, were simply rebreeding their bees each year, offsetting the die-offs with new hives that were more than paid for by the industrial-scale profits they were pulling in.

It's natural, so to speak, to anthropomorphize animals—our whole animation industry is built on it, for starters. But there is something strange, even fatalistic, about such vain beings as ourselves identifying this strongly with creatures who operate so entirely without free will and individual autonomy that many experts in the field aren't sure whether we should think of the bee or the colony as the organism. In my own reporting about colony collapse, bee lovers kept telling me it was an appreciation for the great spectacle of bee civilization that was behind this outpouring of concern for their well-being. But I couldn't help wondering if it wasn't almost the opposite quality that gave colony collapse the force of fable: the complete powerlessness of individuals facing down inevitable, civilization-scale suicide. It's not just bee rapture, after all: we see visions of our own world being wiped out in the mysterious deaths brought about by Ebola, bird flu, and other pandemics; in anxiety about a robot apocalypse; in ISIS, China, and the Jade Helm exercise in Texas; in runaway inflation that never actually happened in the wake of quantitative easing, or the gold rush such fears spawned, which did. One does not open the Wikipedia page for "Honeybee" expecting an encounter with millenarianism. But the more you read about colony collapse, the more you are filled with a kind of awe for just how much the internet is a divining rod by which we choose to intuit an end of days.

As it turns out, there was no mystery, either, about the bee deaths themselves, which could be explained quite fully by the bees' working conditions: mostly that they were rubbing up against a new breed of insecticide, neonicotinoids, which, as the name suggests, effectively turned all the bees into cigarette fiends. Flying insects might be disappearing because of warming, in other words—that recent study suggested that, already, 75 percent of them may have died, drawing us closer to a world without pollinators, which the researchers called an "ecological Armageddon"—but colony collapse disorder has basically nothing to do with that. And yet still, as recently as 2018, magazines were devoting whole feature articles to the bee fable. Presumably, this is not because people enjoyed being wrong about bees, but because treating any apparent crisis as an allegory was somehow comforting—as though it sequestered the problem in a story whose meaning we controlled.

WHEN BILL MCKIBBEN DECLARED "THE END OF NATURE," IN 1989, he was posing a hyperbolic kind of epistemological riddle: What do you call it, whatever *it* is, when forces of wilderness and weather, of animal kingdoms and plant life, have been so transformed by human activity they are no longer truly "natural"?

The answer came a few decades later with the term "the Anthropocene," which was coined in the spirit of environmental alarm and suggested a much messier and more unstable state than "end." Environmentalists, outdoorspeople, nature lovers, and romantics of various stripes—there are many who would mourn the end of nature. But there are literally billions who will shortly be terrified by the forces unleashed by the Anthropocene. In much of the world, they already are, in the form of lethal close-to-annual heat waves in the Middle East and South Asia, and in the ever-present threat of flood, like those that hit Kerala in 2018 and killed hundreds. The floods hardly made a mark in the United States and Europe, where consumers of news have been trained over decades to see disasters like these as tragic, yes, but also as an inevitable feature of underdevelopment—and therefore both "natural" and distant.

The arrival of this scale of climate suffering in the modern West will be one of the great and terrible stories of the coming decades. There, at least, we've long thought that modernity had paved over nature, completely, factory by factory and strip mall by strip mall. Proponents of solar geoengineering want to take on the sky next, not just to stabilize the planet's temperature but possibly to create "designer climates," localized to very particular needs—saving this reef ecosystem, preserving that breadbasket. Conceivably those climates could get considerably more micro, down to particular farms or soccer stadiums or beach resorts.

These interventions, should they ever become feasible, are decades away, at least. But even rapid and quotidian-seeming projects will leave a profoundly different imprint on the shape of the world. In the nineteenth century, the built environment of the most advanced countries reflected the prerogatives of industry—think of railroad tracks laid across whole continents to move coal. In the twentieth century, those same environments were made to reflect the needs of capital—think of global urbanization agglomerating labor supply for a new service economy. In the twenty-first century, they will reflect the demands of the climate crisis: seawalls, carbon-capture plantations, state-sized solar arrays. The claims of eminent domain made on behalf of climate change will no longer play like government overreach, though they will still surely inspire NIMBY backlash—even in a time of climate crisis, progressives will find ways to look out for number one.

We are already living within a deformed environment—indeed, quite deformed. In its swaggering twentieth century, the United States built two states of paradise: Florida, out of dismal swamp, and Southern California, out of desert. By 2100, neither will endure as Edenic postcards.

That we reengineered the natural world so sufficiently to close the book on an entire geological era—that is the major lesson of the Anthropocene. The scale of that transformation remains astonishing, even to those of us who were raised amidst it and took all of its imperious values for granted. Twenty-two percent of the earth's landmass was altered by humans just between 1992 and 2015. Ninety-six percent of the world's mammals, by weight, are now humans and their livestock; just four percent are wild. We have simply crowded—or bullied, or brutalized—every other species into retreat, near-extinction, or worse. E. O. Wilson thinks the era might be better called the Eremocine—the age of loneliness.

But global warming carries a message more concerning still: that we didn't defeat the environment at all. There was no final conquest, no dominion established. In fact, the opposite: Whatever it means for the other animals on the planet, with global warming we have unwittingly claimed ownership of a system beyond our ability to control or tame in any day-to-day way. But more than that: with our continued activity, we have rendered that system only more out of control. Nature is both over, as in "past," and all around us, indeed overwhelming us and punishing us—this is the major lesson of climate change, which it teaches us almost daily. And if global warming continues on anything like its present track, it will come to shape everything we do on the planet, from agriculture to human migration to business and mental health, transforming not just our relationship to nature but to politics and to history, and proving a knowledge system as total as "modernity."

SCIENTISTS HAVE KNOWN THIS FOR A WHILE. BUT THEY HAVE not often talked like it.

For decades now, there have been few things with a worse reputation than "alarmism" among those studying climate change. For a concerned class, this was somewhat strange; you don't typically hear from public health experts about the need for circumspection in describing the risks of carcinogens, for instance. James Hansen, who first testified before Congress about global warming in 1988, has named the phenomenon "scientific reticence," and in 2007 chastised his colleagues for editing their own observations so conscientiously that they failed to communicate how dire the threat really was. That tendency has metastasized over time, ironically as the news from research grew bleaker, so that for a long time each major publication would be attended by a cloud of commentary debating its precise calibration of perspective and tone—with many of those articles seen to lack an even balance between bad news and optimism, and labeled "fatalistic." Some were derided as "climate porn."

The terms are slippery, like any good insult, but served to circumscribe the scope of "reasonable" perspectives on climate. Which is why scientific reticence is another reason we don't see the threat so clearly—the experts signaling strongly that it is irresponsible to communicate openly about the more worrisome possibilities for global warming, as though they didn't trust the world with the information they themselves had, or at least didn't trust the public to interpret it and respond properly. Whatever that means: it has now been thirty years since Hansen's first testimony and the establishment of the IPCC, and climate concern has traversed small peaks and small valleys but never meaningfully jumped upward. In terms of public response, the results are even more dismal. Within the United States, climate denial took over one of the two major parties and essentially vetoed major legislative action. Abroad, we have had a series of high-profile conferences, treaties, and accords, but they increasingly look like so many acts of climate kabuki; emissions are still growing, unabated.

But scientific reticence is also perfectly reasonable, in its way, a river of rhetorical caution with many tributaries. The first is temperamental: climate scientists are scientists first, self-selected and then trained for perspicacity. The second is experiential: many of them have now done battle, in the United States particularly and sometimes for decades, with the forces of climate denial, who capitalize on any overstatement or erroneous prediction as proof of illegitimacy or bad faith; this makes climate scientists more cautious, and understandably so. Unfortunately, worrying so much about erring on the side of excessive alarm has meant they have erred, so routinely it became a kind of professional principle, on the side of excessive caution—which is, effectively, the side of complacency.

There was also a kind of personal wisdom in scientific reticence, as politically backward as it may seem to keep the scariest implications of new research from the public. As part-time advocates, scientists have also watched their colleagues and collaborators pass through many dark nights of the soul, and typically despaired themselves as well, about the coming storm of climate change and just how little the world is doing to combat it. As a result, they were especially worried about burnout, and the possibility that honest storytelling about climate could tip so many people into despondency that the effort to avert a crisis would burn itself out. And in generalizing from that experience, they pointed to a selection of social science suggesting that "hope" can be more motivating than "fear"—without acknowledging that alarm is not the same as fatalism, that hope does not demand silence about scarier challenges, and that fear can motivate, too. That was the finding of a 2017 *Nature* paper surveying the full breadth of the academic literature: that despite a strong consensus among climate scientists about "hope" and "fear" and what qualifies as responsible storytelling, there is no single way to best tell the story of climate change, no single rhetorical approach likely to work on a given audience, and none too dangerous to try. Any story that sticks is a good one.

In 2018, scientists began embracing fear, when the IPCC released a dramatic, alarmist report illustrating just how much worse climate change would be at 2 degrees of warming compared with 1.5: tens of millions more exposed to deadly heat waves, water shortages, and flooding. The research summarized in the report was not new, and temperatures beyond 2 degrees were not even covered. But though it did not address any of the scarier possibilities for warming, the report did offer a new form of permission, of sanction, to the world's scientists. The thing that was new was the message: *It is okay, finally, to freak out.* It is almost hard to imagine, in its aftermath, anything but a new torrent of panic, issuing forth from scientists finally emboldened to scream as they wish to.

But that prior caution was understandable. Scientists spent decades presenting the unambiguous data, demonstrating to anyone who would listen just what kind of crisis will come for the planet if nothing is done, and then watched, year after year, as nothing was done. It should not be altogether surprising that they returned again and again to the communications greenroom, scratching their heads about rhetorical strategy and "messaging." If only they were in charge, they would know exactly what to do, and there would be no need to panic. So why wouldn't anyone listen to them? It had to be the rhetoric. What other explanation could there be?

Crisis
Capitalism

THE SCROLL OF COGNITIVE BIASES IDENTIFIED BY BEHAV-
ioral psychologists and fellow travelers over the last half century is, like a social media feed, apparently infinite, and every single
one distorts and distends our perception of a changing climate—a
threat as imminent and immediate as the approach of a predator,
but viewed always through a bell jar.

There is, to start with, *anchoring*, which explains how we build
mental models around as few as one or two initial examples, no
matter how unrepresentative—in the case of global warming, the
world we know today, which is reassuringly temperate. There is
also the *ambiguity effect*, which suggests that most people are so
uncomfortable contemplating uncertainty, they will accept lesser
outcomes in a bargain to avoid dealing with it. In theory, with climate, uncertainty should be an argument for action—much of the
ambiguity arises from the range of possible human inputs, a quite

concrete prompt we choose to process instead as a riddle, which discourages us.

There is *anthropocentric thinking,* by which we build our view of the universe outward from our own experience, a reflexive tendency that some especially ruthless environmentalists have derided as "human supremacy" and that surely shapes our ability to apprehend genuinely existential threats to the species—a shortcoming many climate scientists have mocked: "The planet will survive," they say; "it's the humans that may not."

There is *automation bias,* which describes a preference for algorithmic and other kinds of nonhuman decision making, and also applies to our generations-long deference to market forces as something like an infallible, or at least an unbeatable, overseer. In the case of climate, this has meant trusting that economic systems unencumbered by regulation or restriction would solve the problem of global warming as naturally, as surely as they had solved the problems of pollution, inequality, justice, and conflict.

These biases are drawn only from the *A* volume of the literature—and are just a sampling of that volume. Among the most destructive effects that appear later in the behavioral economics library are these: *the bystander effect,* or our tendency to wait for others to act rather than acting ourselves; *confirmation bias,* by which we seek evidence for what we already understand to be true, such as the promise that human life will endure, rather than endure the cognitive pain of reconceptualizing our world; the *default effect,* or tendency to choose the present option over alternatives, which is related to the *status quo bias,* or preference for things as they are, however bad that is, and to the *endowment effect,* or instinct to demand more to give up something we have than we actually value it (or had paid to acquire or establish it). We have an *illusion of control,* the behavioral economists tell us, and also suffer from *overconfidence* and an *optimism bias.* We also have a *pessimism bias,* not that it compensates—instead it pushes us to see challenges as predetermined defeats and to hear alarm, perhaps especially on climate, as cries of fatalism. The opposite of a cognitive bias, in other words, is not clear thinking but another cognitive bias. We can't see anything but through cataracts of self-deception.

Many of these insights may feel as intuitive and familiar as folk wisdom, which in some cases they are, dressed up in academic language. Behavioral economics is unusual as a contrarian intellectual movement in that it overturns beliefs—namely, in the perfectly rational human actor—that perhaps only its proponents ever truly believed, and maybe even only as economics undergraduates. But altogether the field is not merely a revision to existing economics. It is a thoroughgoing contradiction of the central proposition of its parent discipline, indeed to the whole rationalist self-image of the modern West as it emerged out of the universities of—in what can only be coincidence—the early industrial period. That is, a map of human reason as an awkward kluge, blindly self-regarding and self-defeating, curiously effective at some things and maddeningly incompetent when it comes to others; compromised and misguided and tattered. How did we ever put a man on the moon?

That climate change demands expertise, and faith in it, at precisely the moment when public confidence in expertise is collapsing, is another of its historical ironies. That climate change touches each of these biases is not a curiosity, or a coincidence, or an anomaly. It is a mark of just how big it is, and how much about human life it touches—which is to say, nearly everything.

You might begin the B volume with BIGNESS—that the scope of the climate threat is so large, and its menace so intense, we reflexively avert our eyes, as we would with the sun.

Bigness as an excuse for complacency will be familiar to anyone who has listened in on an undergraduate debate about capitalism. The size of the problem, its all-encompassing quality, the apparent lack of readymade alternatives, and the enticement of fugitive benefits—these were the building blocks of a decades-long subliminal argument, directed at the increasingly disgruntled professional middle classes of the wealthy West, who on another planet might have formed the intellectual vanguard of a movement against endless financialization and unencumbered markets. "It is easier to imagine the end of the world than to imagine the end of capitalism," the literary critic Fredric Jameson has written, attributing the phrase, coyly, to "someone" who "once said it." That someone might say, today, "Why choose?"

When it comes to authority and responsibility, scale and perspective often befuddle us—we may be unable to recognize which matryoshka doll nests inside the other, or on whose display shelf the whole set sits. Big things make us feel small, and rather powerless, even if we are nominally "in charge." In the modern age, at least, there is also the related tendency to view large human systems, like the internet or industrial economy, as more unassailable, even more un-intervenable, than natural systems, like climate, that literally enclose us. This is how renovating capitalism so that it doesn't reward fossil fuel extraction can seem unlikelier than suspending sulfur in the air to dye the sky red and cool the planet off by a degree or two. To some, even ending trillions in fossil fuel subsidies sounds harder to pull off than deploying technologies to suck carbon out of the air everywhere on Earth.

This is a kind of Frankenstein problem, and relates to widespread fears of artificial intelligence: we are more intimidated by the monsters we create than those we inherit. Sitting at computers in air-conditioned rooms reading dispatches in the science section of the newspaper, we feel illogically in control of natural ecosystems; we expect we should be able to protect the dwindling population of an endangered species, and preserve their habitat, should we choose to, and that we should be able to manage an abundant water supply, rather than see it wasted on the way to human mouths— again, should we choose to. We feel less that way about the internet, which seems beyond our control though we designed and built it, and quite recently; still less about global warming, which we extend each day, each minute, by our actions. And the perceptual size of market capitalism has been a kind of obstacle to its critics for at least a generation, when it came to seem even to those attuned to its failings to be perhaps too big to fail.

It does not quite seem that way now, standing in the long shadow of the financial crisis and watching global warming beginning to darken the horizon. And yet, perhaps in part because we see the way that perspectives on climate change map neatly onto existing and familiar perspectives on capitalism—from burn-it-all-down leftists to naively optimistic and blinkered technocrats to rent-seeking, kleptocratic, growth-is-the-only-value conservatives—we tend to think of climate as somehow being contained within, or governed by, capitalism. In fact, it is endangered by it.

THAT WESTERN CAPITALISM MAY OWE ITS DOMINANCE TO THE power of fossil fuels is not anything like consensus economic wisdom, but it also isn't just a pet theory of the socialist Left. It was the core claim of Kenneth Pomeranz's *The Great Divergence*, probably the single most conventionally esteemed account of just how it was that Europe, long effectively a provincial backwater to the empires of China, India, and the Middle East, separated itself so dramatically from the rest of the world in the nineteenth century. To the big question of "Why Europe?" *The Great Divergence* offers something almost as simple as a one-word answer: coal.

As an account of industrial history, the reductionist story implied by "fossil capitalism"—that what we conceive as the modern economy is really a system powered by fossil fuels—is in ways persuasive but also incomplete; of course there is more to the network that gives us a whole yogurt aisle in the supermarket than the simple burning of oil. (Though maybe less "more" than you'd think.) But as a picture of just how deeply entangled the two forces remain, and how the fate of each defines the fate of the other, the term promises to be a very useful shorthand. And raises the question, now merely rhetorical on parts of the Left: Can capitalism survive climate change?

The question is a prism, spitting out different answers to

different ranges of the political spectrum, and where you fall on that range probably reflects what you mean by "capitalism." Global warming could cultivate emergent forms of eco-socialism on one end of the spectrum, and could also conceivably produce a collapse of faith in anything but the market, on the other. Trade will surely endure, perhaps even thrive, as indeed it did before capitalism—individuals making trades and exchanges outside a single totalizing system to organize the activity. Rent-seeking, too, will continue, with those who can scrambling to accumulate whatever advantages they can buy—the incentive only increasing in a world more barren of resources, and more mournful of recent apparent abundance, now disappeared.

This last is more or less the model that Naomi Klein memorably sketched out in *The Shock Doctrine,* in which she documents just how monolithically the forces of capital respond to crises of any kind—by demanding more space, power, and autonomy for capital. The book is not primarily about the response of financial interests to climate disasters—it focuses more on political collapse and crises of the technocrats' own making. But it does give a very clear account of what kind of strategy to expect from the world's money elite in a time of rolling ecological crisis. More recently, Klein has offered the island of Puerto Rico, still reeling from Hurricane Maria, as a case study, even beyond its unfortunate spot in the path of climate-change-fueled hurricanes. Here is an island endowed with abundant green energy nevertheless importing all its oil, and an agricultural paradise nevertheless importing all its food, importing both from a quasi-colonial mainland power that sees it merely as a market. That mainland power has effectively turned over the government of the island, down to its power company, to a select board of bondholders whose interest is in the repayment of debt.

It is hard to imagine a better illustration of the empire of capital in a time of climate change. And it is not merely rhetorical. In 2017, just after the storm, Solomon Hsiang and Trevor Houser calculated that, all on its own, Maria could cut Puerto Rican incomes by 21 percent over the next fifteen years, and that the economy of the island could take twenty-six years to return even to the level it enjoyed just before the storm—a level, Klein reminds us, already strained. This did not prompt a dramatic expansion of social spending or the extension of a Marshall Plan across the Caribbean; instead, Donald Trump tossed a few rolls of paper towels to the citizens of San Juan, then left them to plead with the outsiders who now controlled the public coffers for mercy, which did not come. The echo of financial crisis is unmistakable, as Hsiang and Houser note, suggesting such crises may offer the best conceptual model for the punishments of climate change. "For Puerto Rico," they write, "Maria could be as economically costly as the 1997 Asian financial crisis was to Indonesia and Thailand and more than twice as damaging as the 1994 Peso Crisis was to Mexico."

How well will the shock doctrine be sustained through a new climatic regime, one that assaults the economies of the world with extreme weather and natural disaster at an entirely unprecedented rate and—just in the diminishing downtime between hurricanes and floods and heat waves and droughts—also threatens to devastate agricultural yields and cripple worker productivity? It is an open question, as are all those having to do with human response to global warming in the present and future. But here, too, even relatively fractional adjustments to the West's basic orientation toward business and financial capitalism are likely to arrive like earthquakes, so much has that orientation produced the culture's collective sense of what is thinkable and what is not.

One possibility is that the scramble for shrinking profits by the powerful will only intensify, a further self-entrenchment of the rule of capital; this is the outcome you might extrapolate from a consideration of the last several decades. But over those decades, capitalists could still count as a public-relations ally the promise of rising-tide growth. In fact, despite our many and divergent varieties of markets, that promise has served as something like the basic ideological infrastructure of the world since at least 1989—and it is no coincidence that carbon emissions have exploded since the end of the Cold War.

Climate change will accelerate two trends already undermining that promise of growth: first, by producing a global economic stagnation that will play, in some areas, like a breathtaking and permanent recession; and second, by punishing the poor much more dramatically than the rich, both globally and within particular polities, showcasing an increasingly stark income inequality, unconscionable already to more and more. In an economic future doubly mangled by those forces, the near-monopoly on social power presently enjoyed by the world's very wealthy will likely have much more to answer for, to say the least.

And how might it answer? Beyond new Social Darwinist appeals to unequal outcomes as "fair" ones, an already familiar one-percenter worldview, the force of capital may find itself with very little to say. The market has justified inequality for generations by pointing to opportunity and invoking the mantra of new prosperity, which it promised would benefit all. This was probably always less credible as a truth claim than it was as propaganda, and, as the Great Recession and the deeply unequal recovery that followed showed unmistakably, income gains in the world's advanced capitalistic countries have gone, for several decades now, almost entirely to the very wealthiest. That this itself represents a crisis of the entire system is clear not just from the raging populism, on both left and right, sweeping Europe and the United States in the aftermath of the crash, but by skepticism and lacerating self-doubt beaming out from the highest free-market citadels. In 2016, the IMF published an article titled "Neoliberalism: Oversold?"—the IMF. And Paul Romer, later the chief economist of the World Bank, proposed that macroeconomics, the "science" of capitalism, was something like a fantasy field, equivalent to string theory, that no longer had any legitimate claim to describing the workings of the real economy accurately. In 2018, Romer won the Nobel Prize. He shared it with William Nordhaus, who pioneered the study of the economic impact of climate change. An economist, Nordhaus favors a carbon tax, but a low one—his "optimal" carbon price still allows for 3.5 degrees Celsius of warming.

At present, the economic impacts of climate change are relatively light: in the United States, in 2017, the estimated cost was $306 billion. The heavier impacts await us. And if, in the past, the promise of growth has been the justification for inequality, injustice, and exploitation, it will have many more wounds to salve in the near climate future: disaster, drought, famine, war, global refugeeism and the political disarray it unleashes. And, as a salve, climate change promises almost no global growth; in much of the world hit hardest, in fact, negative growth.

To the extent that we tend to believe today in human resiliency against such disasters, it is a legacy of several hundred years of industrial affluence produced by our exploitation of fossil fuels. Medieval kings did not believe they could grow their way out of plague, or famine, and those who lived in the shadow of Krakatoa or Vesuvius did not blithely assume they could endure volcanic eruption. But the downward revision of expectations for the future may be still more important than diminished prosperity in the present. And if what you mean by "capitalism" is not just the operation of market forces but the religion of free trade as a just and even perfect social system, you have to expect, at the very least, that a major reformation is coming. The predictions of economic hardship, remember, are enormous—$551 trillion in damages at just 3.7 degrees of warming, 23 percent of potential global income lost, under business-as-usual conditions, by 2100. That is an impact much more severe than the Great Depression; it would be ten times as deep as the more recent Great Recession, which still so rattles us. And it would not be temporary. It is hard to imagine any system surviving that kind of decline intact, no matter how "big."

IF CAPITALISM DOES ENDURE, WHO WILL PAY?

Already, in the United States, courts are awash in a wave of lawsuits aimed at extracting climate damages—a bold gambit, given that most of the impacts they enumerate have yet to arrive. The most high-profile are the torts brought against oil companies by crusading attorneys general—public health claims, more or less, put forward by the public or at least in its name, against companies known to have engaged in disinformation and political-influence campaigns. This is the first vector of climate liability: against the corporations that have profited.

Another kind of charge is made in *Juliana v. the United States*, also known as "Kids vs. Climate," an ingenious equal-protection lawsuit alleging that in failing to take action against global warming, the federal government effectively shifted many decades' worth of environmental costs onto today's young—an inspiring claim, in its way, made by a group of minors on behalf of their entire generation and those that will follow, against the governments their parents and grandparents voted into office. This is the second vector of climate liability: against the generations that have profited.

But there is also a third vector, yet to be litigated in any more formal setting than the conference rooms in which the Paris accords were negotiated: against the nations that have profited from burning fossil fuels, in some cases to the tune of whole empires. This is an especially electric vector because it is the descendants of the subjects of those empires who will bear the bluntest climate trauma—which is what has already inspired the political outrage organized under the banner of "climate justice."

How will those claims play out? A range of scenarios is possible, having to do mostly with what human choices and commitments are made over the next decades. Exploitative empires have collapsed before into relatively peaceful rapprochements, retributive energies muffled by the cushions of reparations, repatriation, truth, and reconciliation. And that could emerge as the dominant approach to climate suffering—a cooperative support network erected in the spirit of mea culpa. But there has been little acknowledgment yet that the wealthy nations of the West owe any climate debt to the poor nations who will suffer most from warming. And that suffering, and the exploitation it expresses, may also prove too gruesome a prompt for high-minded cooperation between nations, many of which could instead look away or retreat into denial.

We do not yet know, of course, just how much suffering global warming will inflict, but the scale of devastation could make that debt enormous, by any measure—larger, conceivably, than any historical debt owed one country or one people by another, almost none of which are ever properly repaid.

If that seems like excessive hyperbole, consider that the British Empire was conjured out of the smoke of fossil fuels and that, today, thanks to that smoke, the marshland of Bangladesh is poised to drown and the cities of India to cook within just the span of a single lifetime. In the twentieth century, the United States did not establish such explicit political dominion, but the global empire it presided over nevertheless transformed many of the nations of the Middle East into oil-pipeline client states—nations now scorched every summer by heat approaching uninhabitable levels in places, and where temperatures are expected to become so hot in the region's holiest mecca that pilgrimages, once the annual rite of millions of Muslims, will be as lethal as genocide. It would take an exceedingly idealistic worldview to believe that the matter of responsibility for that suffering will not fashion our geopolitics in a time of climate crisis, and the cascading flow of that crisis, should we not dam it first, does not offer much foothold for idealism.

Of course, present political arrangements, not to mention bankruptcy law, will conspire to limit climate liability—for oil companies, for governments, for nations. These arrangements may buckle and fall—under force of political pressure and even insurrection—which would have the perhaps unintended effect of clearing from the stage all of the most obvious villains and their guardians, leaving no easy marks to which we might apportion blame and expect commensurate payback. At that point, the matter of blame could become an especially potent and indiscriminate political munition—residual climate rage.

IF WE DO SUCCEED, AND PULL UP SHORT OF TWO OR EVEN THREE degrees, the bigger bill will come due not in the name of liability but in the form of adaptation and mitigation—that is, the cost of building and then administering whatever systems we improvise to undo the damage a century of imperious industrial capitalism has wrought across the only planet on which we all can live.

The cost is large: a decarbonized economy, a perfectly renewable energy system, a reimagined system of agriculture, and perhaps even a meatless planet. In 2018, the IPCC compared the necessary transformation to the mobilization of World War II, but global. It took New York City forty-five years to build three new stops on a single subway line; the threat of catastrophic climate change means we need to entirely rebuild the world's infrastructure in considerably less time.

This is one reason a single-shot cure-all offers an undeniable appeal—which brings us back to that magic phrase, "negative emissions." Neither negative-emissions method—"natural" approaches involving revitalized forests and new agricultural practices, technological ones that would deploy machines to remove carbon from the atmosphere—requires wholesale transformation of the global economy as it is presently constituted. Which is perhaps why negative emissions, once a last-ditch, if-all-else-fails strategy, have recently been built into all conventional climate-action goals. Of 400 IPCC emissions models that land us below two degrees Celsius, 344 feature negative emissions, most of them significantly. Unfortunately, negative emissions are also, at this point, almost entirely theoretical. Neither method has yet been demonstrated to actually work at anything like the necessary scale, but the natural approach, though adored by environmentalists, faces much stiffer obstacles: one researcher suggested that, to succeed, it would require a third of the world's farmable land; another suggested that, depending on exactly how the system was designed and deployed, it might have the opposite of its intended effect, not subtracting carbon from the atmosphere but adding it.

The carbon capture path, which would blanket the planet in anti-industrial plants out of a cyberpunk dream, seems, by contrast, more inviting. To begin with, we already have the technology, though it is expensive. The devices, Wallace Smith Broecker is fond of saying, have about the same mechanical complexity as a car, and cost about as much—roughly $30,000 each. To merely match the amount of carbon we are presently emitting into the atmosphere, Broecker calculates, would require 100 million of them. This would merely buy us some time—at a cost of $30 trillion, or about 40 percent of global GDP. To reduce the level of carbon in the atmosphere just by a few parts per million—which would buy us a little more time, matching not just our present emissions but our likely level a few years down the road—would take 500 million of these devices. To reduce the level of carbon by 20 parts per million per year, he calculates, would require 1 billion of them. This would immediately pull us back from the threshold, even buy us some more time of carbon growth—which is an argument you hear against it from some corners of the environmental Left. But it would cost, you may already have calculated, $300 trillion—or nearly four times total global GDP.

These prices will likely fall, but only as emissions and atmospheric carbon continue to rise. In 2018, a paper by David Keith demonstrated a method for removing carbon at a cost perhaps as low as $94 per ton—which would make the cost of neutralizing our 32 gigatons of annual global emissions about $3 trillion. If that sounds intimidating, keep in mind, estimates for the total global fossil fuel subsidies paid out each year run as high as $5 trillion. In 2017, the same year the United States pulled out of the Paris Agreement, the country also approved a $2.3 trillion tax cut—primarily for the country's richest, who demanded relief.

The Church of
Technology

S HOULD ANYTHING SAVE US, IT WILL BE TECHNOLOGY. BUT you need more than tautologics to save the planet, and, especially within the futurist fraternity of Silicon Valley, technologists have little more than fairy tales to offer. Over the last decade, consumer adoration has anointed those founders and venture capitalists something like shamans, Ouija-boarding their way toward blueprints for the world's future. But conspicuously few of them seem meaningfully concerned about climate change. Instead, they make parsimonious investments in green energy (Bill Gates aside) and fewer still philanthropic payouts (Bill Gates again aside), and often express the perspective, outlined by Eric Schmidt, that climate change has already been solved, in the sense that a solution has been made inevitable by the speed of technological change—or even by the introduction of a particular self-advancing technology, namely machine intelligence, or AI.

Blind faith is one way of describing this worldview, though many in Silicon Valley regard machine intelligence with blind terror. Another way of looking at it is that the world's futurists have come to regard technology as a superstructure within which all other problems, and their solutions, are contained. From that perspective, the only threat to technology must come from technology, which is perhaps why so many in Silicon Valley seem less concerned with runaway climate change than they are with runaway artificial intelligence: the only fearsome power they are likely to take seriously is the one they themselves have unleashed. It is a strange evolutionary stage for a worldview midwifed into being, in the permanent counterculture of the Bay Area, by Stewart Brand's nature-hacking bible, *Whole Earth Catalog*. And it may help explain why social media executives were so slow to process the threat that real-world politics posed to their platforms; and perhaps also why, as the science fiction writer Ted Chiang has suggested, Silicon Valley's fear of future artificial-intelligence overlords sounds suspiciously like an unknowingly lacerating self-portrait, panic about a way of doing business embodied by the tech titans themselves:

> Consider: Who pursues their goals with monomaniacal focus, oblivious to the possibility of negative consequences? Who adopts a scorched-earth approach to increasing market share? This hypothetical strawberry-picking AI does what every tech startup wishes it could do—grows at an exponential rate and destroys its competitors until it's achieved an absolute monopoly. The idea of superintelligence is such a poorly defined notion that one could envision it taking almost any form with equal justification: a benevolent genie that solves all the world's problems, or a mathematician that spends all its time proving theorems so abstract that humans can't even understand them. But when Silicon Valley tries to imagine superintelligence, what it comes up with is no-holds-barred capitalism.

SOMETIMES IT CAN BE HARD TO HOLD MORE THAN ONE extinction-level threat in your head at once. Nick Bostrom, the pioneering philosopher of AI, has managed it. In an influential 2002 paper taxonomizing what he called "existential risks," he outlined twenty-three of them—risks "where an adverse outcome would either annihilate Earth-originating intelligent life or permanently and drastically curtail its potential."

Bostrom is not a lone doomsday intellectual but one of the leading thinkers currently strategizing ways of corralling, or at any rate conceptualizing, what they consider the species-sized threat from an out-of-control AI. But he does include climate change on his big-picture risk list. He puts it in the subcategory "Bangs," which he defines as the possibility that "earth-originating intelligent life goes extinct in relatively sudden disaster resulting from either an accident or a deliberate act of destruction." "Bangs" is the longest of his sub-lists; climate change shares the category with, among others, *Badly programmed superintelligence* and *We're living in a simulation and it gets shut down*.

In his paper, Bostrom also considers the climate-change-adjacent risk of "resource depletion or ecological destruction." He places that threat in his next category, "Crunches," which he describes as an episode after which "the potential of humankind to develop into posthumanity is permanently thwarted although human life continues in some form." His most representative crunch risk is probably *Technological arrest:* "the sheer technological difficulties in making the transition to the posthuman world might turn out to be so great that we never get there." Bostrom's final two categories are "Shrieks," which he defines as the possibility that "some form of posthumanity is attained but it is an extremely narrow band of what is possible and desirable," as in the case of "Take-over by a transcending upload" or "Flawed superintelligence" (as opposed to "Badly programmed superintelligence"); and "Whimpers," which he defines as "a posthuman civilization arises but evolves in a direction that leads gradually but irrevocably to either the complete disappearance of the things we value or to a state where those things are realized to only a minuscule degree of what could have been achieved."

As you may have noticed, although his paper sets out to analyze "human extinction scenarios," none of his threat assessments beyond "Bangs" actually mention "humanity." Instead, they are focused on what Bostrom calls "posthumanity" and others often call "transhumanism"—the possibility that technology may quickly carry us across a threshold into a new state of being, so divergent from the one we know today that we would be forced to consider it a true rupture in the evolutionary line. For some, this is simply a vision of nanobots swimming through our bloodstreams, filtering toxins and screening for tumors; for others, it is a vision of human life extracted from tangible reality and uploaded entirely to computers. You may notice here an echo of the Anthropocene. In this vision, though, humans aren't burdened with environmental wreckage and the problem of navigating it; instead, we simply achieve a technological escape velocity.

It is hard to know just how seriously to take these visions, though they are close to universal among the Bay Area's futurist vanguard, who have succeeded the NASAs and the Bell Labs of the last century as architects of our imagined future—and who differ among themselves primarily in their assessments of just how long it will take for all this to come to pass. Peter Thiel may complain about the pace of technological change, but maybe he's doing so because he's worried it won't outpace ecological and political devastation. He's still investing in dubious eternal-youth programs and buying up land in New Zealand (where he might ride out social collapse on the civilization scale). Y Combinator's Sam Altman, who has distinguished himself as a kind of tech philanthropist with a small universal-basic-income pilot project and recently announced a call for geoengineering proposals he might invest in, has reportedly made a down payment on a brain-upload program that would extract his mind from this world. It's a project in which he is also an investor, naturally.

For Bostrom, the very purpose of "humanity" is so transparently to engineer a "posthumanity" that he can use the second term as a synonym for the first. This is not an oversight but the key to his appeal in Silicon Valley: the belief that the grandest task before technologists is not to engineer prosperity and well-being for humanity but to build a kind of portal through which we might pass into another, possibly eternal kind of existence, a technological rapture in which conceivably many—the billions lacking access to broadband, to begin with—would be left behind. It would be very hard, after all, to upload your brain to the cloud when you're buying pay-as-you-go data by the SIM card.

The world that would be left behind is the one being presently pummeled by climate change. And Bostrom isn't alone, of course, in identifying that risk as species-wide. There are the thousands, perhaps hundreds of thousands, of scientists now seeming to scream daily, with each extreme-weather event and new research paper, for the attention of lay readers; and no more hysterical a figure than Barack Obama was fond of using the phrase "existential threat." And yet it is perhaps a sign of our culture's heliotropism toward technology that aside perhaps from proposals to colonize other planets, and visions of technology liberating humans from most biological or environmental needs, we have not yet developed anything close to a religion of meaning around climate change that might comfort us, or give us purpose, in the face of possible annihilation.

OF COURSE, THOSE ARE RELIGIOUS FANTASIES: TO ESCAPE THE body and transcend the world.

The first is almost a caricature of privileged thinking, and that it should have entered the dream lives of a new billionaire caste was probably close to inevitable. The second seems like a strategic response to climate panic—securing a backup ecosystem to hedge against the possibility of collapse here—which is precisely as it has been described by its advocates.

But the solution is not a rational one. Climate change does threaten the very basis of life on this planet, but a dramatically degraded environment here will still be much, much closer to livability than anything we might be able to hack out of the dry red soil of Mars. Even in summer, at the equator of that planet, nighttime temperatures are a hundred degrees Fahrenheit below zero; there is no water on its surface, and no plant life. Conceivably, given sufficient funding, a small enclosed colony could be built there, or on another planet; but the costs would be so much higher than for an equivalent artificial ecosystem on Earth, and therefore the scale so much more limited, that anyone proposing space travel as a solution to global warming must be suffering from their own climate delusion. To imagine such a colony could offer material prosperity as abundant as tech plutocrats enjoy in Atherton is to live even more deeply in the narcissism of that delusion—as though it were only as difficult to smuggle luxury to Mars as to Burning Man.

The faith takes a different form among the laity, unable to afford that ticket into space. But articles of faith are offered, considerately, at different price points: smartphones, streaming services, rideshares, and the internet itself, more or less free. And each glimmers with some promise of escape from the struggles and strife of a degraded world.

In "An Account of My Hut," a memoir of Bay Area house-hunting and climate-apocalypse-watching in the 2017 California wildfire season—which was also the season of Hurricanes Harvey and Irma and Maria—Christina Nichol describes a conversation with a young family member who works in tech, to whom she tried to describe the unprecedentedness of the threat from climate change, unsuccessfully. "Why worry?" he replies.

> "Technology will take care of everything. If the Earth goes, we'll just live in spaceships. We'll have 3D printers to print our food. We'll be eating lab meat. One cow will feed us all. We'll just rearrange atoms to create water or oxygen. Elon Musk."

Elon Musk—it's not the name of a man but a species-scale survival strategy. Nichol answers, "But I don't *want* to live in a spaceship."

He looked genuinely surprised. In his line of work, he'd never met anyone who didn't want to live in a spaceship.

THAT TECHNOLOGY MIGHT LIBERATE US, COLLECTIVELY, FROM the strain of labor and material privation is a dream at least as old as John Maynard Keynes, who predicted his grandchildren would work only fifteen-hour weeks, and yet never ultimately fulfilled. In 1987, the year he won the Nobel Prize, economist Robert Solow famously commented, "You can see the computer age everywhere but in the productivity statistics."

This has been, even more so, the experience of most of those living in the developed world in the decades since—rapid technological change transforming nearly every aspect of everyday life, and yet yielding little or no tangible improvement in any conventional measures of economic well-being. It is probably one explanation for contemporary political discontent—a perception that the world is being almost entirely remade, but in a way that leaves you, as delighted as you may be by Netflix and Amazon and Instagram and Google Maps, more or less exactly where you were before.

The same can be said, believe it or not, for the much-heralded green energy "revolution," which has yielded productivity gains in energy and cost reductions far beyond the predictions of even the most doe-eyed optimists, and yet has not even bent the curve of carbon emissions downward. We are, in other words, billions of dollars and thousands of dramatic breakthroughs later, precisely where we started when hippies were affixing solar panels to their geodesic domes. That is because the market has not responded to these developments by seamlessly retiring dirty energy sources and replacing them with clean ones. It has responded by simply adding the new capacity to the same system.

Over the last twenty-five years, the cost per unit of renewable energy has fallen so far that you can hardly measure the price, today, using the same scales (since just 2009, for instance, solar energy costs have fallen more than 80 percent). Over the same twenty-five years, the proportion of global energy use derived from renewables has not grown an inch. Solar isn't eating away at fossil fuel use, in other words, even slowly; it's just buttressing it. To the market, this is growth; to human civilization, it is almost suicide. We are now burning 80 percent more coal than we were just in the year 2000.

And energy is, actually, the least of it. As the futurist Alex Steffen has incisively put it, in a Twitter performance that functions as a "Powers of Ten" for the climate crisis, the transition from dirty electricity to clean sources is not the whole challenge. It's just the lowest-hanging fruit: "smaller than the challenge of electrifying almost everything that uses power," Steffen says, by which he means anything that runs on much dirtier gas engines. That task, he continues, is smaller than the challenge of reducing energy demand, which is smaller than the challenge of reinventing how goods and services are provided—given that global supply chains are built with dirty infrastructure and labor markets everywhere are still powered by dirty energy. There is also the need to get to zero emissions from all other sources—deforestation, agriculture, livestock, landfills. And the need to protect all human systems from the coming onslaught of natural disasters and extreme weather. And the need to erect a system of global government, or at least international cooperation, to coordinate such a project. All of which is a smaller task, Steffen says, "than the monumental cultural undertaking of imagining together a thriving, dynamic, sustainable future that feels not only possible, but worth fighting for."

On this last point I see things differently—the imagination isn't the hard part, especially for those less informed about the challenges than Steffen is. If we could wish a solution into place by imagination, we'd have solved the problem already. In fact, we *have* imagined the solutions; more than that, we've even developed them, at least in the form of green energy. We just haven't yet discovered the political will, economic might, and cultural flexibility to install and activate them, because doing so requires something a lot bigger, and more concrete, than imagination—it means nothing short of a complete overhaul of the world's energy systems, transportation, infrastructure and industry and agriculture. Not to mention, say, our diets or our taste for Bitcoin. The cryptocurrency now produces as much CO_2 each year as a million transatlantic flights.

WE THINK OF CLIMATE CHANGE AS SLOW, BUT IT IS UNNERVingly fast. We think of the technological change necessary to avert it as fast-arriving, but unfortunately it is deceptively slow—especially judged by just how soon we need it. This is what Bill McKibben means when he says that winning slowly is the same as losing: "If we don't act quickly, and on a global scale, then the problem will literally become insoluble," he writes. "The decisions we make in 2075 won't matter."

Innovation, in many cases, is the easy part. This is what the novelist William Gibson meant when he said, "The future is already here, it just isn't evenly distributed." Gadgets like the iPhone, talismanic for technologists, give a false picture of the pace of adaptation. To a wealthy American or Swede or Japanese, the market penetration may seem total, but more than a decade after its introduction, the device is used by less than 10 percent of the world; for all smartphones, even the "cheap" ones, the number is somewhere between a quarter and a third. Define the technology in even more basic terms, as "cell phones" or "the internet," and you get a timeline to global saturation of at least decades—of which we have two or three, in which to completely eliminate carbon emissions, planetwide. According to the IPCC, we have just twelve years to cut them in half. The longer we wait, the harder it will be. If we had started global decarbonization in 2000, when Al Gore narrowly lost election to the American presidency, we would have had to cut emissions by only about 3 percent per year to stay safely under two degrees of warming. If we start today, when global emissions are still growing, the necessary rate is 10 percent. If we delay another decade, it will require us to cut emissions by 30 percent each year. This is why U.N. Secretary-General António Guterres believes we have only one year to change course and get started.

The scale of the technological transformation required dwarfs any achievement that has emerged from Silicon Valley—in fact dwarfs every technological revolution ever engineered in human history, including electricity and telecommunications and even the invention of agriculture ten thousand years ago. It dwarfs them by definition, because it contains all of them—every single one needs to be replaced at the root, since every single one breathes on carbon, like a ventilator.

To remake each of these systems so that they don't is less like distributing smartphones or floating wifi balloons over Kenya or Puerto Rico, as Google intends to, than like building an interstate highway system or constructing a subway network or a new kind of power grid connected to a new array of energy producers and new kind of energy consumer. In fact, it is not *like* that; it *is* that. All of that and much, much more: intensive infrastructure projects at every level and in every corner of human activity, from new plane fleets to new land use and right down to a new way of making concrete, production of which ranks today as the second most carbon-intensive industry in the world—an industry that is booming, by the way, thanks to China, which recently poured more concrete in three years than the United States used in the entire twentieth century. If the cement industry were a country, it would be the world's third-largest emitter.

In other words, these are infrastructure projects of a scale so far from our experience, in the U.S. at least, that we hardly expect their existing corollaries to ever even be repaired anymore, instead learning to live with potholes and service delays. On top of which, unlike the internet or smartphones, the requisite technologies are not additive but substitutive, or should be, if we have the good sense to actually retire the dirty old varieties. Which means that all of the new alternatives have to face off with the resistance of entrenched corporate interests and the status-quo bias of consumers who are relatively happy with the lives they have today.

Thankfully, the green energy revolution is already, as they say, "under way." In fact, of all the necessary components of this broader, zero-carbon revolution, clean energy is probably farthest along. How far along? In 2003, Ken Caldeira, now of the Carnegie Institution for Science, found that the world would need to add clean power sources equivalent to the full capacity of a nuclear plant every single day between 2000 and 2050 to avoid catastrophic climate change. In 2018, *MIT Technology Review* surveyed our progress; with three decades left to go, the world was on track to complete the necessary energy revolution in four hundred years.

That gap yawns so wide it could swallow whole civilizations, and indeed threatens to. Into it has crawled that dream of carbon capture: if we can't rebuild the entire infrastructure of the modern world in time to save it from self-destruction, perhaps we can at least buy ourselves some time by sucking some of its toxic fumes out of the air. Given the indomitable scale of the conventional approach, and given just how little time left we have to complete it, negative emissions may be, at present, a form of magical thinking for climate. They also seem like a last, best hope. And if they work, carbon capture plants will deliver industrial absolution for industrial sin—and initiate, as a result, a whole new theological romance with the power of machine.

THREADED THROUGH THE REVERIE FOR CARBON CAPTURE IS A fantasy of industrial absolution—that a technology could be almost dreamed into being that could purify the ecological legacy of modernity, even perhaps eliminate its footprint entirely.

The semi-subliminal sales pitch for wind and solar is not dissimilar—clean energy, natural energy, renewable and therefore sustainable energy, inexhaustible, even undiminishable energy, harnessed rather than harvested energy, abundant energy, free energy. Which all sounds quite a lot like nuclear power, at least as it was originally presented and received. Of course, that was back in the 1950s, and it has been decades now since nuclear was seen as a path to energy salvation rather than, as it invariably is today, through the specter of metaphysical contagion.

It was not always this way. In his 1953 "Atoms for Peace" speech before the United Nations, Dwight Eisenhower outlined the terms of a standing-offer arms trade that was also a moral bargain: as a reward to any nation disavowing the pursuit of nuclear weapons, and as a kind of penance for having developed the horrible technology in the first place, the United States would offer aid in the form of nuclear energy, which it would also be cultivating at home.

For a speech delivered by a president who was also a military man, it is a remarkably lyrical lament that is also a peacetime call-to-arms—in fact, it evokes in a modern reader quite beautifully the threat from climate change. After briefly describing the dramatic expansion of the capacity of the American nuclear fleet, which had in the eight years since the war grown twenty-five times more powerful and plainly terrified him, and then what it meant for the United States to have gained Soviet Russia as a nuclear rival, Eisenhower continues:

> To stop there would be to accept helplessly the probability of civilization destroyed, the annihilation of the irreplaceable heritage of mankind handed down to us from generation to generation, and the condemnation of mankind to begin all over again the age-old struggle upward from savagery towards decency, and right, and justice. Surely no sane member of the human race could discover victory in such desolation. Could anyone wish his name to be coupled by history with such human degradation and destruction? Occasional pages of history do record the faces of the "great destroyers," but the whole book of history reveals mankind's never-ending quest for peace and mankind's God-given capacity to build.

It has been at least a generation since Americans might have casually read "mankind's God-given capacity to build" as a reference to nuclear power—a generation since the world stopped believing nuclear power was, in an environmental sense, "free," and started thinking of it in terms of nuclear war, meltdown, mutation, and cancer. That we remember the names of power-plant disasters is a sign of just how scarred we feel by them: Three Mile Island, Chernobyl, Fukushima.

But the scars are almost phantom ones, given the casualty numbers. The death toll of the incident at Three Mile Island is in some dispute, as many activists believe the true impact of radiation was suppressed—perhaps a reasonable belief, since the official account insists on no adverse health impacts at all. But the most pedigreed research suggests the meltdown increased cancer risk, within a ten-mile radius, by less than one-tenth of 1 percent. For Chernobyl, the official death count is 47, though some estimates run higher—even as high as 4,000. For Fukushima, according to a United Nations report, "no discernible increased incidence of radiation-related health effects are expected among exposed members of the public or their descendants." Had none of the 100,000 living in the evacuation zone ever left, perhaps a few hundred might have ultimately died of cancers related to the radiation.

Any number of dead is a tragedy, but more than 10,000 people die each day, globally, from the small-particulate pollution produced by burning carbon. This is not even broaching the subject of warming and its impacts. A rule change to pollution standards for coal producers, proposed by Trump's EPA in 2018, would kill an additional 1,400 Americans annually, the agency itself acknowledged; globally, pollution kills as many as nine million each year.

We live with that pollution, and with those death tolls, and hardly notice them; the curving concrete towers of nuclear plants, by contrast, stand astride the horizon like Chekhov's proverbial gun. Today, despite a variety of projects aimed at producing cheap nuclear energy, the price of new plants remains high enough that it is hard to make a persuasive argument that more "green" investment be directed toward them rather than installations of wind and solar. But the case for decommissioning and dismantling existing plants is considerably weaker, and yet that is exactly what is happening—

from the United States, where both Three Mile Island and Indian Point are being closed down, to Germany, where so much nuclear power has recently been retired that the country is growing its carbon emissions despite a state-of-the-world green energy program. For this, Angela Merkel has been called the "Climate Chancellor."

THE CONTAMINATIONIST VIEW OF NUCLEAR POWER IS A MIS-guided climate parable, arising nevertheless from a perceptive environmentalist perspective—that the healthy, clean natural world is made toxic by the intrusions and interventions of human industry. But the main lesson from the church of technology runs in the other direction, instructing us in subtle and not-so-subtle ways to regard the world beyond our phones as less real, less urgent, and less meaningful than the worlds made available to us through those screens, which happen to be worlds protected from climate devastation. As Andreas Malm has wondered, "How many will play augmented reality games on a planet that is six degrees warmer?" The poet and musician Kate Tempest puts it more brinily: "Staring into the screen so we don't have to see the planet die."

Presumably, you can already feel this transformation underfoot, in your own life—scrolling through photos of your baby when your actual baby is right in front of you, reading trivial Twitter threads while your spouse is speaking. In Silicon Valley, even tech critics tend to see the problem as a form of addiction; but, like all addictions, it expresses a value judgment, if one that makes the un-addicted uncomfortable—in this case, that we find the world of our screens more rewarding, or safer, in ways so hard to justify and explain that there isn't really a word for it other than "preferable." This preference is much more likely to grow than shrink, which may seem like cultural devolution, perhaps especially to temperamental declinists. It could conceivably also be a psychologically useful coping mechanism for living, still within the consumptive bourgeois tradition, in a dramatically degraded natural world. A generation from now, god help us, tech addiction may even look "adaptive."

Politics of Consumption

J ust before dawn on April 14, 2018, a Saturday, a sixty-year-old man walked into Prospect Park in Brooklyn, gave himself a shower of gasoline, and lit himself on fire. Beside the body, near a circular patch of grass blackened by the flames, lay a note, handwritten: "I am David Buckel and I just killed myself by fire as a protest suicide," it read. "I apologize to you for the mess." It was a small mess; he had arranged a ring of soil to prevent the fire from spreading too far.

In a longer letter, typed, which he had also sent to the city's newspapers, Buckel elaborated. "Most humans on the planet now breathe air made unhealthy by fossil fuels, and many die early deaths as a result—my early death by fossil fuel reflects what we are doing to ourselves. . . . Pollution ravages our planet," he wrote. "Our present grows more desperate, our future needs more than what we've been doing."

AMERICANS KNOW POLITICAL SUICIDE BY SELF-IMMOLATION from the Vietnam era, when Thích Quảng Đức, a Buddhist monk repurposing a spiritual tradition of self-purification for contemporary protest, burned himself to death in Saigon. A few years later, the thirty-one-year-old Quaker Norman Morrison was inspired to do the same, outside the Pentagon, his one-year-old daughter beside him. One week after that, twenty-two-year-old Roger Allen LaPorte, a former seminarian and Catholic Worker, lit himself aflame outside the United Nations. We don't like to think about it, but the tradition continues. In the United States, there have been six protests by self-immolation since 2014; in China, the gesture is even more common, particularly by opponents of the country's Tibet policy, with twelve in the last three months of 2011 and twenty in the first three months of 2012 alone. And of course the self-immolation of a Tunisian fruit vendor ignited the Arab Spring.

Buckel was a later-life environmental activist. He'd spent most of his career as a prominent gay-rights litigator, and his notes expressed two clear convictions: that the natural world had been made sick by industrial activity, and that much more than the average passersby in Prospect Park could appreciate must be done to halt, and ideally reverse, the damage. In the days after his suicide, it was the first of these which attracted the most attention—his death treated as an alarm, or a bellwether, marking some amorphous shift, perhaps in the health of the planet but certainly in the average Brooklynite's perception of it. The second insight is more challenging—that the climate crisis demands political commitment well beyond the easy engagement of rhetorical sympathies, comfortable partisan tribalism, and ethical consumption.

It is a common charge against liberal environmentalists that they live hypocritically—eating meat, flying, and voting liberal without yet having purchased Teslas. But among the woke Left the inverted charge is just as often true: we navigate by a North Star of politics through our diets, our friendships, even our consumption of pop culture, but rarely make meaningful political noise about those causes that run against our own self-interest or sense of self as special—indeed enlightened. And so, in the coming years, divestment is likely to be just the first salvo in a moral arms race between universities, municipalities, and nations. Cities will compete to be the first to ban cars, to paint every single roof white, to produce all the agriculture eaten by residents in vertical farms that don't require post-harvest transportation by automobile, railroad, or airplane. But liberal NIMBYism will still strut, too, as it did in 2018, when American voters in deep-blue Washington state rejected a carbon tax at the ballot box, and the worst French protests since the quasi-revolution of 1968 raged against a proposed gasoline tax. On perhaps no issue more than climate is that liberal posture of well-off enlightenment a defensive gesture: almost regardless of your politics or your consumption choices, the wealthier you are, the larger your carbon footprint.

But when critics of Al Gore compare his electricity use to that of the average Ugandan, they are not ultimately highlighting conspicuous and hypocritical personal consumption, however they mean to disparage him. Instead, they are calling attention to the structure of a political and economic order that not only permits the disparity but feeds and profits from it—this is what Thomas Piketty calls the "apparatus of justification." And it justifies quite a lot. If the world's most conspicuous emitters, the top 10 percent, reduced their emissions to only the E.U. average, total global emissions would fall by 35 percent. We won't get there through the dietary choices of individuals, but through policy changes. In an age of personal politics, hypocrisy can look like a cardinal sin; but it can also articulate a public aspiration. Eating organic is nice, in other words, but if your goal is to save the climate your vote is much more important. Politics is a moral multiplier. And a perception of worldly sickness uncomplemented by political commitment gives us only "wellness."

IT CAN BE HARD TO TAKE WELLNESS SERIOUSLY AS A MOVE-ment, at first, which may be why it has been the subject of so much derision over the past few years—SoulCycle, Goop, Moon Juice. But however manipulated by marketing consultants, and however dubious its claims to healthfulness, wellness also gives a clear name and shape to a growing perception even, or especially, among those wealthy enough to be insulated from the early assaults of climate change: that the contemporary world is toxic, and that to endure or thrive within it requires extraordinary measures of self-regulation and self-purification.

What has been called the "new New Age" arises from a similar intuition—that meditation, ayahuasca trips, crystals and Burning Man and microdosed LSD are all pathways to a world beckoning as purer, cleaner, more sustaining, and perhaps above all else, more whole. This purity arena is likely to expand, perhaps dramatically, as the climate continues to careen toward visible degradation—and consumers respond by trying to extract themselves from the sludge of the world however they can. It should not be a surprise to discover, in next year's supermarket aisles, alongside labels for "organic" and "free range," some food described as "carbon-free." GMOs aren't a sign of a sick planet but a possible partial solution to the coming crisis of agriculture; nuclear power the same for energy. But both have already become nearly as off-putting as carcinogens to the purity-minded, who are growing in number and channeling more and more ecological anxiety along the way.

That anxiety is coherent, even rational, at a time when it has been revealed that many American brand-name foods made from oats, including Cheerios and Quaker Oats, contain the pesticide Roundup, which has been linked with cancer, and when the National Weather Service issues elaborate guidance about which commonly available face masks can, and which cannot, protect you against the wildfire smoke engulfing nearly all of North America. It is only intuitive, in other words, that impulses toward purity represent growth areas of our culture, destined to distend further inward from the cultural periphery as apocalyptic ecological anxiety grows, too.

But conscious consumption and wellness are both cop-outs, arising from that basic promise extended by neoliberalism: that consumer choices can be a substitute for political action, advertising not just political identity but political virtue; that the mutual end-goal of market and political forces should be the effective retirement of contentious politics at the hand of market consensus, which would displace ideological dispute; and that, in the meantime, in the supermarket aisle or department store, one can do good for the world simply by buying well.

THE TERM "NEOLIBERALISM" HAS BEEN A SWEAR WORD, ON THE Left, only since the Great Recession. Before then it was, most of the time, mere description: of the growing power of markets, particularly financial markets, in the liberal democracies of the West over the second half of the twentieth century; and of the hardening centrist consensus within those countries committed to spreading that power, in the form of privatization, deregulation, corporate-friendly tax policy, and the promotion of free trade.

This program was sold, for fifty years, on the promise of growth—and not just growth for some. In this way, it was a sort of total political philosophy, extending a single, simple ideological tarpaulin so far and wide that it enclosed the earth like a rubbery blanket of greenhouse gas.

It was total in other ways as well, unable to adjust to meaningfully discriminate between experiences as divergent as post-crash England and post-Maria Puerto Rico, or to concede its own short-comings and paradoxes and blind spots, proposing instead only more neoliberalism. This is how the forces that unleashed climate change—namely, "the unchecked wisdom of the market"—were nevertheless presented as the forces that would save the planet from its ravages. It is how "philanthrocapitalism," which seeks profits alongside human benefits, has replaced the loss-leader model of moral philanthropy among the very rich; how the winners of our increasingly winner-take-all tournament economy use philanthropy to buttress their own status; how "effective altruism," which measures even not-for-profit charity by metrics of return borrowed from finance, has transformed the culture of giving well beyond the billionaire class; and how the "moral economy," a rhetorical wedge that once expressed a radical critique of capitalism, became the calling card of do-gooder capitalists like Bill Gates. It is also, on the other end of the pecking order, how struggling citizens are asked to be entrepreneurs, indeed to demonstrate their value as citizens with the hard work of entrepreneurship, in an exhausting social system defined above all else by relentless competition.

That is the critique from the Left, at least—and it is, in its way, inarguably true. But by laundering all conflict and competition through the market, neoliberalism also proffered a new model of doing business, so to speak, on the world stage—one that didn't emerge from, or point toward, endless nation-state rivalry.

One should not confuse correlation with causation, especially since there was so much tumult coming out of World War II that it is hard to isolate the single cause of just about anything. But the international cooperative order that has since presided, establishing or at least emerging in parallel with relative peace and abundant prosperity, is very neatly historically coincident with the reign of globalization and the empire of financial capital we now group together as neoliberalism. And if one were inclined to confuse correlation with causation, there is a quite intuitive and plausible theory connecting them. Markets may be problematic, shall we say, but they also value security and stability and, all else being equal, reliable economic growth. In the form of that growth, neoliberalism promised a reward for cooperation, effectively transforming, at least in theory, what had once been seen as zero-sum competitions into positive-sum collaborations.

Neoliberalism never made good on that bargain, as the financial crisis finally revealed. Which has left the rhetorical banner of

an ever-expanding, ever-enriching society of affluence—and a political economy oriented toward the same goal—considerably tattered. Those continuing to hold it aloft are much wobblier at the
knee than seemed credible to imagine just a decade or two ago, like
athletes showing themselves suddenly far past their prime. Global
warming promises another blow, possibly a lethal one. If Bangladesh floods and Russia profits, the result will not be good for the
cause of neoliberalism—and arguably worse still for the cause of
liberal internationalism, which has always been its aide-de-camp.

What kinds of politics are likely to evolve after the promise of
growth recedes? A whole pantheon of possibilities floats before us,
including that new trade deals are built on the moral infrastructure
of climate change, with commerce contingent on emissions cuts
and sanctions a punishment for squirrelly carbon behavior; or that
a new global legal regime emerges, supplementing or perhaps even
supplanting the central principle of human rights that has presided
globally, at least in theory, since the end of World War II. But
neoliberalism was sold on the proposition of positive-sum cooperation of all kinds, and the term itself suggests its natural successor
regime: zero-sum politics. Today, we don't even have to gaze into
the future, or trust that it will be deformed by climate change, to
see what that would look like. In the form of tribalism at home and
nationalism abroad and terrorism flaming out from the tinder of
failed states, that future is here, at least in preview, already. Now
we just wait for the storms.

IF NEOLIBERALISM IS THE GOD THAT FAILED ON CLIMATE
change, what juvenile gods will it spawn? This is the question taken
up by Geoff Mann and Joel Wainwright in *Climate Leviathan: A
Political Theory of Our Planetary Future,* in which they repurpose
Thomas Hobbes to sketch out what they see as the likeliest political form to evolve from the crisis of warming and the pummeling
of its impacts.

In his *Leviathan,* Hobbes narrated a false history of political consent to illustrate what he saw as the fundamental bargain of state power: the people giving up their liberty for the protection offered by a king. Global warming suggests the same bargain to would-be authoritarians: in a newly dangerous world, citizens will trade liberties for security and stability and some insurance against climate deprivation, ushering into being, Mann and Wainwright say, a new form of sovereignty to do battle against the new threat from the natural world. This new sovereignty will be not national but planetary—the only power that could plausibly answer a planetary threat.

Mann and Wainwright are leftists, and their book is in part a call-to-arms, but the planetary sovereign the world is likeliest to turn to, they say with regret, is the one that sold us climate change in the first place—that is, neoliberalism. In fact, a neoliberalism beyond neoliberalism, a true world-state concerned close-to-exclusively with the flow of capital—a preoccupation that may poorly equip it to deal with the damages and degradations of climate change, but at no real cost to its authority. This is the "Climate Leviathan" of the title, though the authors do not believe its success is inevitable. In fact, they see three variations as also possible. Altogether, the four categories make up a climate-future matrix, plotted along the axes of relative faith in capitalism (on the one hand) and degree of support for nation-state sovereignty (on the other).

"Climate Leviathan" is the quadrant defined by a positive relationship to capitalism and a negative perspective on national sovereignty. Something like our current situation they call the "Climate Behemoth" outcome, defined by mutual support for capitalism and for the nation-state: capitalism overruns the world's borders to address the planetary crisis while protecting its own interests.

The next they call "Climate Mao," a system defined by putatively benevolent but authoritarian and anti-capitalist leaders, exercising their authority within the borders of nations as they exist today.

The last quadrant: capitalistic nations conduct haphazard climate diplomacy—an international system negatively disposed toward both capitalism and the sovereignty of nation-states. This system would define itself as a guarantor of stability and security— ensuring at least a subsistence-level distribution of resources, protecting against the ravages of extreme climate events, and policing the inevitable outbreaks of conflict over the now-more-precious commodities of food, water, land. It would also wipe out entirely the borders between nations, recognizing only its own sovereignty and power. They call this possibility "Climate X," and express great hope for it: a global alliance operating in the name of a common humanity, rather than in the interests of capital or nations. But there is a dark version as well—it is how you might get a planetary dictator in the shape of a mafia boss, and global governance not on the do-gooder model but as a straight-up protection racket.

In theory, at least. Already, it's fair to say, we have at least two Climate Mao leaders out there, and both are imperfect avatars of the archetype: Xi Jinping and Vladimir Putin, neither of whom is anti-capitalist so much as state capitalist. They also hold very different perspectives on the climate future and how to reckon with it, which suggests another variable, beyond form of government: climate ideology. This is how Angela Merkel and Donald Trump, both operating within the "Climate Behemoth" system, can nevertheless seem so many worlds apart—though Germany's slow walk on coal suggests there may not be full solar systems between them.

With China and Russia, the ideological contrast is clearer. Putin, the commandant of a petro-state that also happens to be, given its geography, one of the few nations on Earth likely to benefit from continued warming, sees basically no benefit to constraining carbon emissions or greening the economy—Russia's or the world's. Xi, now the leader-for-life of the planet's rising superpower, seems to feel mutual obligations to the country's growing prosperity and to the health and security of its people—of whom, it's worth remembering, it has so many.

In the wake of Trump, China has become a much more

emphatic—or at least louder—green energy leader. But the incentives do not necessarily suggest it will make good on that rhetoric. In 2018, an illuminating study was published comparing how much a country was likely to be burdened by the economic impacts of climate change to its responsibility for global warming, measured by carbon emissions. The fate of India showcased the moral logic of climate change at its most grotesque: expected to be, by far, the world's most hard-hit country, shouldering nearly twice as much of the burden as the next nation, India's share of climate burden was four times as high as its share of climate guilt. China is in the opposite situation, its share of guilt four times as high as its share of the burden. Which, unfortunately, means it may be tempted to slow-walk its green energy revolution. The United States, the study found, presented a case of eerie karmic balance: its expected climate damages matching almost precisely its share of global carbon emissions. Not to say either share is small; in fact, of all the nations in the world, the U.S. was predicted to be hit second hardest.

For decades, the rise of China has been an anxious prophecy invoked so regularly, and so prematurely, that Westerners, Americans especially, could be forgiven for thinking it was a case of the empire who cried wolf—an expression of Western self-doubt, more a premonition of collapse than a well-founded prediction of what new power might arise, and when. But on the matter of climate change, China does hold nearly all the cards. To the extent the world as a whole needs a stable climate to endure or thrive, its fate will be determined much more by the carbon trajectories of the developing world than by the course of the United States and Europe, where emissions have already flattened out and will likely begin their decline soon—though how dramatic a decline, and how soon, is very much up in the air. And although what's called "carbon outsourcing" means that a large slice of China's emissions is produced manufacturing goods to be consumed by Americans and Europeans. Whose responsibility are those gigatons of carbon? It may not much longer be merely a rhetorical question, if the Paris accords yield to a more rigorous global carbon governance structure, as they were intended to, and add, along the way, a proper enforcement mechanism, military or otherwise.

How and how fast China manages its own transition from industrial to postindustrial economy, how and how fast it "greens" the industry that remains, how and how fast it remodels agricultural practices and diet, how and how fast it steers the consumer preferences of its booming middle and upper classes away from carbon intensity—these are not the only things that will determine the climate shape of the twenty-first century. The courses taken by India and the rest of South Asia, Nigeria and the rest of sub-Saharan Africa, matter enormously. But China is, at present, the largest of those nations, and by far the wealthiest and most powerful. Through its Belt and Road Initiative, the country has already positioned itself as a major provider, in some cases the major provider, of the infrastructure of industry, energy, and transportation in much of the rest of the developing world. And it is relatively easy to imagine, at the end of a Chinese century, an intuitive global consensus solidifying—that the country with the world's largest economy (therefore most responsible for the energy output of the planet) and the most people (therefore most responsible for the public health and well-being of humanity) should have something more than narrowly national powers over the climate policy of the rest of the "community of nations," who would fall into line behind it.

ALL OF THESE SCENARIOS, EVEN THE BLEAKEST, PRESUME SOME new political equilibrium. There is also, of course, the possibility of disequilibrium—or what you would normally call "disorder" and "conflict." This is the analysis put forward by Harald Welzer, in *Climate Wars,* which predicts a "renaissance" of violent conflict in the decades to come. His evocative subtitle is *What People Will Be Killed For in the 21st Century.*

Already, in local spheres, political collapse is a quite common outcome of climate crisis—we just call it "civil war." And we tend to analyze it ideologically—as we did in Darfur, in Syria, in

Yemen. Those kinds of collapses are likely to remain technically "local" rather than truly "global," though in a time of climate crisis they would have an easier time metastasizing beyond old borders than they have in the recent past. In other words, a completely *Mad Max* world is not around the bend, since even catastrophic climate change won't undermine all political power—in fact, it will produce some winners, relatively speaking. Some of them with quite large armies and rapidly expanding surveillance states—China now pulls criminals out of pop concerts with facial recognition software and deploys domestic-spy drones indistinguishable from birds. This is not an aspiring empire likely to tolerate no-man's-lands within its sphere.

Mad Max regions elsewhere are another matter. In certain ways they are already here, where "here" is parts of Somalia or Iraq or South Sudan at various points in the last decade, including points when the planet's geopolitics seemed, at a glance from Los Angeles or London, stable. The idea of a "global order" has always been something of a fiction, or at least an aspiration, even as the joined forces of liberal internationalism, globalization, and American hegemony inched us toward it over the last century. Very probably, over the next century, climate change will reverse that course.

History After Progress

T HAT HISTORY IS A STORY THAT MOVES IN ONE DIRECTION is among the most unshakable creeds of the modern West—having survived, often only slightly modified, the counterarguments made over centuries by genocides and gulags, famines and epidemics and global conflagrations, producing death tolls in the tens of millions. The grip of this narrative is so tight on political imaginations that grotesque injustices and inequities, racial and otherwise, are often invoked not as reasons to doubt the arc of history but to be reminded of its shape—perhaps we shouldn't be quite so agitated about such problems, in other words, since history is "moving in the right direction" and the forces of progress are, to indulge the mixed metaphor, "on the right side of history." On what side is climate change?

Its own side—its own tide. There is no good thing in the world that will be made more abundant, or spread more widely, by global warming. The list of the bad things that will proliferate is innumerable. And already, in this age of nascent ecological crisis, you can read a whole new literature of deep skepticism—proposing not only that history can move in reverse, but that the entire project of human settlement and civilization, which we know as "history" and which has given us climate change, has been, in fact, a jet stream backward. As climate horrors accumulate, this anti-progressive perspective is sure to blossom.

Some Cassandras are already here. In *Sapiens*, his alien's-eye-view account of the rise of human civilization, the historian Yuval Noah Harari argues that this rise is best understood as a succession of myths, beginning with the one that the invention of farming, in what is often called the Neolithic Revolution, amounted to progress ("We did not domesticate wheat. It domesticated us," as he pithily put it). In *Against the Grain*, the political scientist and anthropologist of anarchy James C. Scott gives a more pointed critique of the same period: wheat cultivation, he argues, is responsible for the arrival of what we now understand as state power, and, with it, bureaucracy and oppression and inequality. These are no longer outlier accounts of what you may have learned about in middle school as the Agricultural Revolution, which you probably were taught marked the real beginning of history. Modern humans have been around for 200,000 years, but farming for only about 12,000—an innovation that ended hunting and gathering, bringing about cities and political structures, and with them what we now think of as "civilization." But even Jared Diamond—whose *Guns, Germs, and Steel* gave an ecological and geographical account of the rise of the industrial West, and whose *Collapse* is a kind of forerunner text for this recent wave of reconsiderations—has called the Neolithic Revolution "the worst mistake in the history of the human race."

The argument does not even rely on anything that followed later: industrialization, fossil fuels, or the damage they now threaten to unleash on the planet and the fragile civilization briefly erected on its slippery surface. Instead, the case against civilization, this new class of skeptics says, can be made much more directly as a case against farming: the sedentary life agriculture produced eventually led to denser settlements, but populations didn't expand for millennia afterward, the potential growth from farming offset by new levels of disease and warfare. This was not a brief, painful interlude, through which humans passed into a new time of abundance, but a story of strife that continued for a very long time, indeed to this day. We are still, now, in much of the world, shorter, sicker, and dying younger than our hunter-gatherer forebears, who were also, by the way, much better custodians of the planet on which we all live. And they watched over it for much longer—nearly all of those 200,000 years. That epic era once derided as "prehistory" accounts for about 95 percent of human history. For nearly all of that time, humans traversed the planet but left no meaningful mark. Which makes the history of mark-making—the entire history of civilization, the entire history we know as history—look less like an inevitable crescendo than like an anomaly, or blip. And makes industrialization and economic growth, the two forces that really gave the modern world the hurtling sensation of material progress, a blip inside a blip. A blip inside a blip that has brought us to the brink of a never-ending climate catastrophe.

James Scott comes to this subject as a radical anti-statist, toward the end of a long career producing genuinely scintillating works of academic dissidence with titles like *The Art of Not Being Governed, Domination and the Arts of Resistance,* and *Two Cheers for Anarchism.* Harari's is a stranger approach, but also more telling—a from-the-roots reconsideration of our collective faith in human progress, put forward and gobbled up in the midst of an ecological crisis of our own making. Harari has spoken movingly of the way his own coming out, as a gay man, has shaped his skepticism about human metanarratives as pervasive as heterosexuality and progress; and, though trained as a military historian, he has arrived in the spotlight of popular acclaim, praised by Bill Gates and Barack Obama and Mark Zuckerberg, as a sort of expositor of myth. The central exposition is this: society is and always has been bound together by collective fictions, no less now than in earlier eras, with values like progress and rationality taking the place once held by religion and superstition. Harari is a historian, but his worldview grafts the pretense of science onto the philosophical skepticism familiar from contrarians as diverse as David Hume and John Gray. You could also name the whole line of French theorists, from Lyotard to Foucault and beyond.

"The story that has ruled our world in the past few decades is what we might call the Liberal Story," Harari wrote in 2016, a month before the election of Donald Trump, in an essay that both basically predicted Trump's election and outlined what it would mean to the world's collective faith in the establishment. "It was a simple and attractive tale, but it is now collapsing, and so far no new story has emerged to fill the vacuum."

IF YOU STRIP OUT THE PERCEPTION OF PROGRESS FROM HIS-tory, what is left?

From here, it is hard, if not impossible, to see clearly what will emerge from the clouds of uncertainty around global warming—what forms we allow climate change to take, let alone what those forms will do to us. But it will not take a worst-case warming to deliver ravages dramatic enough to shake the casual sense that as time marches forward, life improves ineluctably. Those ravages are likely to begin arriving quickly: new coastlines retreated from drowned cities; destabilized societies disgorging millions of refugees into neighboring ones already feeling the pinch of resource depletion; the last several hundred years, which many in the West saw as a simple line of progress and growing prosperity, rendered instead as a prelude to mass climate suffering. Exactly how we regard the shape of history in a time of climate change will be shaped by how much we do to avert that change and how much we let it remodel everything about our lives. In the meantime, possibilities fan out as extravagantly as the paint chips on a color wheel.

We still don't know all that much about how humans before the arrival of agriculture, statehood, and "civilization" regarded the course of history—though it was a favorite pastime of early modern philosophers to imagine the inner lives of precivilized people, from "nasty, brutish, and short" to idyllic, carefree, unencumbered.

Another perspective, which offers another model of history, is the cyclical one: familiar from the harvest calendar, the Stoic Greek theory of ekpyrosis and the Chinese "dynastic cycle," and appropriated for the modern era by thinkers as seemingly teleological as Friedrich Nietzsche, who made the cycles of time a moral parable with his "eternal recurrence"; Albert Einstein, who considered the possibility of a "cyclic" model of the universe; Arthur Schlesinger, who saw American history as alternating periods of "public purpose" and "private interest"; and Paul Michael Kennedy, in his circumspect history lesson for the end of the Cold War, *The Rise and Fall of the Great Powers*. Perhaps Americans today see history as progressive only because we were raised in the time of its empire, having more or less borrowed the British perspective from the time of theirs.

But climate change isn't likely to deliver a neat or complete return to a cyclical view of history, at least in the premodern sense—in part because there will be nothing neat, at all, about the era ushered in by warming. The likelier outcome is a much messier perspective, with teleology demoted from its position as an organizing, unifying theory, and, in its place, contradictory narratives running uncorralled, like animals unleashed from a cage and moving in all directions at once. But if the planet reaches three or four or five degrees of warming, the world will be convulsed with human suffering at such a scale—so many million refugees, half again as many wars, droughts and famines, and economic growth made impossible on so much of the planet—that its citizens will have difficulty regarding the recent past as a course of progress or even a phase in a cycle, or in fact anything but a true and substantial reversal.

The possibility that our grandchildren could be living forever among the ruins of a much wealthier and more peaceful world seems almost inconceivable from the vantage of the present day, so much do we still live within the propaganda of human progress and generational improvement. But of course it was a relatively common feature of human history before the advent of industrialization. It was the experience of the Egyptians after the invasion of the Sea Peoples and the Incas after Pizarro, the Mesopotamians after the Akkadian Empire, and the Chinese after the Tang Dynasty. It was—so famously that it grew into caricature, which then spawned decades of rhetorical critique—the experience of Europeans after the fall of Rome. But in this case, the dark ages would arrive within one generation of the light—close enough to touch, and share stories, and blame.

THIS IS WHAT IS MEANT WHEN CLIMATE CHANGE IS DESCRIBED as a revenge of time. "Man-made weather is never made in the present," Andreas Malm writes in *The Progress of This Storm*, his powerful sketch of a political theory for a time of climate change. "Global warming is a result of actions in the past."

It's a tidy formulation, and one that vividly illustrates both the scale and the scope of the problem, which appears as the product of several long centuries of carbon-burning that also produced most of what we think of today as the comforting features of modern life. In that way, climate change does make us all prisoners of the Industrial Revolution, and suggests a carceral model of history— progress arrested by the consequences of past behavior. But while the climate crisis was engineered in the past, it was mostly in the recent past; and the degree to which it transforms the world of our grandchildren is being decided not in nineteenth-century Manchester but today and in the decades ahead.

Disorientingly, climate change will also send us hurtling forward into an uncharted future—so long forward, if it proceeds unchecked, and into such a distant future, that we can hardly imagine the scale. This is not the "techno-shock" first experienced by Victorians encountering an accelerating pace of progress and feeling overwhelmed by just how much was changing within a single lifetime—though we are now acquainting ourselves with that kind of change, as well. It is more like the overwhelming awe felt by those naturalists contemplating the ancient-beyond-ancient historical grandeur of the earth, and calling it deep time.

But climate change inverts the perspective—giving us not a deep time of permanence but a deep time of cascading, disorienting change, so deep that it mocks any pretense of permanence on the planet. Pleasure districts like Miami Beach, built just decades ago, will disappear, as will many of the military installations erected around the world since World War II to defend and secure the wealth that gave rise to them. Much older cities, like Amsterdam, are also under threat from flooding, with extraordinary infrastructure needed already today to keep them above water, infrastructure unavailable to defend the temples and villages of Bangladesh. Farmlands that had produced the same strains of grain or grapes for centuries or more will adapt, if they are lucky, to entirely new crops; in Sicily, the breadbasket of the ancient world, farmers are already turning to tropical fruits. Arctic ice that formed over millions of years will be unleashed as water, literally changing the face of the planet and remodeling shipping routes responsible for the very idea of globalization. And mass migrations will sever communities numbering in the millions—even tens of millions—from their ancestral homelands, which will disappear forever.

Just how long the ecosystems of Earth will be thrown into flux and disarray from anthropogenic climate change also depends on how much more of that change we choose to engineer—and perhaps how much we can manage to undo. But warming at the level necessary to fully melt ice sheets and glaciers and elevate sea level by several hundred feet promises to initiate rolling, radically transformative changes on a timescale measured not in decades or centuries or even millennia, but in the millions of years. Alongside that timeline, the entire lifespan of human civilization is rendered, effectively, an afterthought; and the much longer span of climate change becomes eternity.

Ethics at the End of the World

T HE TWIN TOWNS OF SAN IGNACIO AND SANTA ELENA, Belize, are fifty miles from the coast and 250 feet above sea level, but the alarmist climatologist Guy McPherson did not move there—to a farm in the jungle that surrounds the towns—in fear of water. Other things will get him first, he says; he's given up hope of surviving climate change, and believes the rest of us should, too. Humans will be extinct within ten years, he tells me by Skype; when I ask his partner, Pauline, if she feels the same way, she laughs. "I'd say ten months." This was two years ago.

McPherson began his career as a conservation biologist at the University of Arizona, where, he mentions several times, he was tenured at twenty-nine; and where, he also says several times, he was surveilled by what he calls the "Deep State" beginning in 1996; and also where, in 2009, he was forced out of his department by a new chair. He had already been working on a homestead in New Mexico—a compromise location with his former wife— and moved in 2016 to the Central American jungle, to live with

Pauline and practice polyamory on another homestead called Stardust Sanctuary Farm.

Over the last decade, mostly via YouTube, McPherson has acquired what Bill McKibben calls, in his understated way, "a following." These days, McPherson travels a bit, giving lectures on "near-term human extinction," a term he is proud to have coined and which he abbreviates NTHE; but increasingly he has turned his attention to running workshops on what we should do with the knowledge that the world is ending. The workshops are called "Only Love Remains," and offer what amounts to a kind of post-theological millenarianism, familiar hand-me-down lessons from the old New Age. The meta-lesson is that we should draw roughly the same meaning from an understanding of the imminent death of the species as the Dalai Lama believes we should draw from an understanding of our imminent personal death—namely, compassion, wonderment, and above all, love. You could do worse in choosing three values around which to build an ethical model, and when you squint you can almost see a civics erected out of them. But for those who see the planet as being on the precipice of crisis and biblical tribulation, they also excuse a retreat from politics—indeed from climate, as fully as that might conceivably be achieved—in the name of a slippery hedonistic quietism.

In other words, down to the mustache, McPherson seems like a recognizable off-the-grid figure—a kind it's easy to find a bit suspicious. But why? We have for so long, over decades if not centuries, defined predictions of the collapse of civilization or the end of the world as something close to proof of insanity, and the communities that spring up around them as "cults," that we are now left unable to take any warnings of disaster all that seriously—especially when those raising the alarm are also, themselves, "giving up." There is nothing the modern world abhors like a quitter, but that prejudice will probably not withstand much warming. If the climate crisis unfolds as it is scheduled to, our taboos against doomsaying will fall, as new cults emerge and cultish thinking leeches into sectors of establishment culture. Because while the world will not likely end, and civilization is almost surely more resilient than McPherson believes, the unmistakable degradation of the planet will invariably inspire many more prophets like him, whose calls of imminent environmental apocalypse will start to seem reasonable to many more reasonable people.

That is, in part, because they are not so unreasonable, even today. If you were looking for a primer on the bad news about climate, you could find a worse place to begin than the summary page McPherson keeps on his website, "Nature Bats Last" (currently tagged with this note: "Updated most recently, likely for the final time, 2 August 2016"). It runs sixty-eight printed pages of link-dense paragraphs. Throughout, there are misleading characterizations of serious research, and links to hysterical, uncredentialed blog posts presented as references to solid science. There are simple misunderstandings of things like climate feedback loops, which can worryingly add up but are not "multiplicative," as McPherson says they are; attacks on merely moderate climate groups as politically compromised; and, in the spirit of a kitchen-sink data dump, endorsements of a few observations that have been proven to be bunk (he is very worried, for instance, about those methane "burps of death" going off all at once, a possibility specialists turned against about five years ago). But, even on this fearmongering reading list, there is enough real science to give rise to real alarm: a good summary of the albedo effect, a convenient assemblage of rigorous readings of the Arctic ice sheets, those tea leaves of climate disaster.

Throughout, the intellectual style is paranoid—the impressive mass of data sometimes standing in for, and sometimes obscuring, the skeleton of causal logic that should give the mass a meaningful analytic shape. This kind of reasoning lives abundantly on the internet, feeding our golden age of conspiracy theory, that insatiable beast, which has only just begun to feast on climate. You might know already the shape that thinking takes on the climate-denial end of the political spectrum. But it has also made its mark on the environmentalist fringe, as it did in the person of

John B. McLemore, the charismatic, closeted environmental de-
clinist and self-hating Southerner whose descent into suicide, beset
by planetary panic, was documented on the podcast "S-Town." "I
sometimes call it toxic knowledge," Richard Heinberg of the Post
Carbon Institute, where McLemore was a commenter, has said.
"Once you *know* about overpopulation, overshoot, depletion, climate change, and the dynamics of societal collapse, you can't *un-
know* it, and your every subsequent thought is tinted."

McPHERSON ISN'T ENTIRELY CLEAR HIMSELF ON EXACTLY HOW
all of these problems will bring about extinction—he guesses that
something like a food crisis or financial meltdown will bring down
civilization first, and eventually human life with it. It takes an
apocalyptic imagination to picture that happening just a decade
from now, to be sure. But, given the basic trend lines, it also raises
the question of why the rest of us aren't imagining things more
apocalyptically ourselves.

We surely will, and soon. Already you can see the seedlings of
a great flourishing of climate esoterica in figures like McLemore
and McPherson—one might better say "men," as they nearly all
are—and, beyond them, a whole harvest of writers and thinkers
who seem, in their anticipation of coming disasters, almost to be
cheering for the forces of apocalypse.

In some cases, they are rooting them on quite literally. A few,
like McLemore, are Travis Bickles of climate crisis, hoping to see
a hard rain fall and wash away all the world's scum. But there are
also Pollyannaish connoisseurs of global warming, like ecologist
Chris D. Thomas, who argues that, in fact, in the real-time vacuum
of the sixth mass extinction, "nature is thriving"—inventing new
species, carving new ecological niches. Some technologists and
their fans go further, suggesting we should discard our bias for
the present—even in the attenuated geological sense of the term
"present"—and adopt instead a quasi-Taoist climate sanguinity,
layered over with a futurist cast. As Swedish journalist Torill

Kornfeldt asks in *The Re-Origin of Species,* her book about the race to "de-extinct" creatures like dinosaurs and woolly mammoths: "Why should nature as it is now be of any greater value than the natural world of 10,000 years ago, or the species that will exist 10,000 years from now?"

BUT FOR MOST WHO PERCEIVE AN ALREADY UNFOLDING CLImate crisis and intuit a more complete metamorphosis of the world to come, the vision is a bleak one, often pieced together from perennial eschatological imagery inherited from existing apocalyptic texts like the Book of Revelation, the inescapable sourcebook for Western anxiety about the end of the world. In fact, those ravings, which Yeats more or less translated for a secular audience in "The Second Coming," have so predominated the Western dreamscape—becoming something like the Gnostic wallpaper of our bourgeois inner lives—that we often forget they were originally written as real-time prophecies, visions of what was to come, and what would become of the world, within a single generation.

Probably the most prominent of these new climate Gnostics is the British writer Paul Kingsnorth, the cofounder, public face, and poet laureate of the Dark Mountain Project, a loose renunciation community of disaffected environmentalists that takes its name from the American writer Robinson Jeffers, in particular his 1935 poem "Rearmament," which ends:

> *I would burn my right hand in a slow fire*
> *To change the future . . . I should do foolishly. The beauty of modern*
> *Man is not in the persons but in the*
> *Disastrous rhythm, the heavy and mobile masses, the dance of the*
> *Dream-led masses down the dark mountain.*

Jeffers was, for a time, a literary celebrity in America—a love affair chronicled in the *Los Angeles Times,* a granite home on the

California coast called Tor House and Hawk Tower, which he famously built with his own hands. But he is known today primarily as a prophet of civilizational disavowal, and for the philosophy he bluntly called "inhumanism": the belief, in short, that people were far too concerned with people-ness, and the place of people in the world, rather than the natural majesty of the nonhuman cosmos in which they happened to find themselves. The modern world, he believed, made the problem considerably worse.

Edward Abbey adored Jeffers's work, and Charles Bukowski called him his favorite poet. The great American wilderness photographers—Ansel Adams, Edward Weston—were influenced by him, too; and in *A Secular Age,* the philosopher Charles Taylor identified Jeffers, alongside Nietzsche and Cormac McCarthy, as a significant figure of what he called "immanent anti-humanism." In his most infamous work, "The Double Axe," Jeffers put that worldview in the mouth of a single character, "The Inhumanist," who described "a shifting of emphasis and significance from man to not-man; the rejection of human solipsism and recognition of the transhuman magnificence." This would be a genuine revolution in perspective, he wrote, which "offers a reasonable detachment as a rule of conduct, instead of love, hate, and envy."

That detachment forms the core principle of Dark Mountain— though one might better say "impulse." It will likely animate many more groups of environmental retreatists over the next decades, if global warming makes the broad spectacle of life on Earth increasingly unbearable for some to observe, even through media. "Those who witness extreme social collapse at first hand seldom describe any deep revelation about the truths of human existence," the group's manifesto begins. "What they do mention, if asked, is their surprise at how easy it is to die. The pattern of ordinary life, in which so much stays the same from one day to the next, disguises the fragility of its fabric."

In that manifesto, written by Kingsnorth and Dougald Hine and first published in 2009, the group identifies as its intellectual godfather Joseph Conrad, particularly for the way he skewered the self-serving illusions of European civilization at its industrial-colonial peak. They quote Bertrand Russell recapping Conrad, saying that the author of *Heart of Darkness* and *Lord Jim* "thought of civilized and morally tolerable human life as a dangerous walk on a thin crust of barely cooled lava which at any moment might break and let the unwary sink into fiery depths." It would be a vivid image to brandish in any era, but especially in a time of approaching ecological collapse. "We believe that the roots of these crises lie in the stories we have been telling ourselves," Kingsnorth and Hine write—namely, "the myth of progress, the myth of human centrality, and the myth of our separation from 'nature.'" All, they add, "are more dangerous for the fact that we have forgotten they are myths."

In fact, it is almost hard to think of anything that won't be changed by just the perception of onrushing change, from the way couples contemplate the possibility of children all the way up to political incentive structure. And you don't have to get all the way to human extinction or the collapse of civilization for true nihilism and doomsdayism to flourish—you only have to get far enough from the familiar for a critical mass of charismatic prophets to see a coming collapse. It can be comforting to think that the critical mass is quite large, and that societies won't be upended by nihilism unless nihilism becomes the conventional view of the median citizen. But doom works at the margin, too, eating away at the infrastructure of things like termites or carpenter bees.

In 2012, Kingsnorth published a new manifesto, or pseudo-manifesto, in *Orion*, called "Dark Ecology." In the meantime, he had grown even less hopeful. "Dark Ecology" opens with epigraphs from Leonard Cohen and D. H. Lawrence—"Take the only tree that's left / Stuff it up the hole in your culture" and "Retreat to the desert, and fight," respectively—and really kicks into gear with its second section, which opens: "I've recently been read-

ing the collected writings of Theodore Kaczynski. I'm worried that
it may change my life."

All told, the essay, which inspired an enormous response among
Orion readers, is a kind of argument on behalf of Kaczynski the
pamphleteer against Kaczynski the bomber—whom Kingsnorth
describes not as a nihilist or even a pessimist but an incisive ob-
server whose problem was an excess of optimism, a man too committed to the idea that society could be changed. Kingsnorth is
more of a true Stoic. "And so I ask myself: what, at this moment in
history, would not be a waste of my time?"

He offers five tentative answers. Numbers 2 through 4 are variations on new transcendentalist themes: "preserving nonhuman
life," "getting your hands dirty," and "insisting that nature has
value beyond utility." Numbers 1 and 5 are the more radical ones,
and form a pair: "withdrawing" and "building refuges." The latter is
the more positive imperative, in the sense of being constructive, or
what passes for constructive in a time of collapse: "Can you think,
or act, like the librarian of a monastery through the Dark Ages,
guarding the old books as empires rise and fall outside?"

"Withdrawing" is the darker half of the same admonition:

If you do this, a lot of people will call you a "defeatist" or a
"doomer," or claim you are "burnt out." They will tell you
that you have an obligation to work for climate justice or
world peace or the end of bad things everywhere, and that
"fighting" is always better than "quitting." Ignore them,
and take part in a very ancient practical and spiritual tradition: withdrawing from the fray. Withdraw not with cyni-
cism, but with a questing mind. Withdraw so that you can
allow yourself to sit back quietly and feel, intuit, work out
what is right for you and what nature might need from you.
Withdraw because refusing to help the machine advance—
refusing to tighten the ratchet further—is a deeply moral
position. Withdraw because action is not always more effective than inaction. Withdraw to examine your worldview:

the cosmology, the paradigm, the assumptions, the direction of travel. All real change starts with withdrawal.

It's an ethos, at least. And one with a pedigree. What might read at first like a radical response to a new moment of crisis is in fact a repurposing of the long and many-armed ascetic tradition, stretching from the young Buddha through the pillar saints and beyond. But unlike the conventional version, in which the ascetic impulse carries the seeker away from the pleasures of the world toward spiritual meaning in something like worldly pain, Kingsnorth's withdrawal, like McPherson's, is a retreat from a world convulsed by spiritual pain toward small, earthly consolations. In that way, it is a performance at grand scale of the more general prophylactic reflex we share, almost all of us, toward suffering—which is to say, simply, an aversion. And to what end? It can't possibly be that I feel the anguish of others, and the urgency of action, through the "myth" of civilization alone—can it?

DARK MOUNTAIN IS FRINGE. GUY McPHERSON IS FRINGE. John B. McLemore, too. But one threat of climate catastrophe is that their strains of ecological nihilism might find a home in the host of consensus wisdom—and that their premonitions may seem familiar to you is a sign that some of that anxiety and despair is already leeching into the way so many others think about the future of the world. Online, the climate crisis has given rise to what is called "eco-fascism"—a "by any means necessary" movement that also traffics in white supremacy and prioritizes the climate needs of a particular set. On the left, there is a growing admiration for the climate authoritarianism of Xi Jinping.

In the United States, the go-it-alone impulse to environmentalist separatism has been predominantly the domain of right-wing extremists—Cliven Bundy and his family, for instance, and all the imperious settlers the country has uncomplicatedly mythologized in the centuries since homesteading and range wars. Perhaps in response, liberal environmentalism has grown mostly in a more practical direction, tending toward more engagement rather than the opposite. Or perhaps it just reflects the particular demands of this cause: form a renunciation community and risk having those you've renounced do everything you feared they might, unleashing changes to the planet you are powerless to escape.

But this pragmatism brings its own curiosities—for instance, that many of even those who define themselves as practical technocrats of the environmental center-left believe that what is needed to avert catastrophic climate change is a global mobilization at the scale of World War II. They are right—that is an entirely sober assessment of the size of the problem, which no more alarmist a group than the IPCC endorsed in 2018. But it is also an undertaking of ambitions so inconsistent with the present tense of politics in nearly every corner of the world, that it is hard not to worry what will happen when that mobilization does not happen—to the planet, yes, but also to the political commitments of those most engaged with the problem. Those calling for mass mobilization, starting today and no later, remember—they can be counted as environmental technocrats. To their left are those who see no solution short of political revolution. And even those activists are being crowded for space, these days, by texts of climate alarmism, of which you may even feel the book in your hands is one. That would be fair enough, because I am alarmed.

I am not alone. And how widespread alarm will shape our ethical impulses toward one another, and the politics that emerge from those impulses, is among the more profound questions being posed by the climate to the planet of people it envelops. It is one way to understand why activists in California were so frustrated with their governor, Jerry Brown, even though he established a climate program of surpassing ambition just as he left office—because he didn't act aggressively enough to retire existing fossil fuel capacity. It also helps explain frustration with other leaders, from Justin Trudeau, who has seized the rhetorical mantle of climate action but also approved several new Canadian pipelines, to Angela Merkel, who has overseen an exhilarating expansion of Germany's green energy capacity, but also retired its nuclear power so quickly that some of the slack has been taken up by existing dirty plants. To the average citizen of each of these countries, the criticism may seem extreme, but it arises from a very clearheaded calculus: the world has, at most, about three decades to completely decarbonize before truly devastating climate horrors begin. You can't halfway your way to a solution to a crisis this large.

In the meantime, environmental panic is growing, and so is despair. Over the last several years, as unprecedented weather and unrelenting research have recruited more voices to the army of environmental panic, a dour terminological competition has sprung up among climate writers, aiming to coin new clarifying language—in the mode of Richard Heinberg's "toxic knowledge" or Kris Bartkus's "Malthusian tragic"—to give epistemological shape to the demoralizing, or demoralized, response of the rest of the world. To the environmental indifference expected of modern consumers, the philosopher and activist Wendy Lynne Lee has given the name "eco-nihilism." Stuart Parker's "climate nihilism" is easier on the tongue. Bruno Latour, an instinctive insubordinate, calls the menace of a raging environment fueled by indifferent politics a "climatic regime." We have also "climate fatalism" and "ecocide" and what Sam Kriss and Ellie Mae O'Hagan, making a psychoanalytic argument against the relentless public-facing optimism of environmental advocacy, have called "human futilitarianism":

> The problem, it turns out, is not an overabundance of humans but a dearth of humanity. Climate change and the Anthropocene are the triumph of an undead species, a mindless shuffle toward extinction, but this is only a lopsided imitation of what we really are. This is why political depression is important: zombies don't feel sad, and they certainly don't feel helpless; they just are. Political depression is, at root, the experience of a creature that is being prevented from being itself; for all its crushingness, for all

its feebleness, it's a cry of protest. Yes, political depressives feel as if they don't know how to be human; buried in the despair and self-doubt is an important realization. If humanity is the capacity to act meaningfully within our surroundings, then we are not really, or not yet, human.

The novelist Richard Powers points his finger at a different kind of despair, "species loneliness," which he identifies not as the impression left on us by environmental degradation but what has inspired us, seeing the imprint we are leaving, to nevertheless continue pressing onward: "the sense we're here by ourselves, and there can be no purposeful act except to gratify ourselves." As though initiating a more accommodationist wing of Dark Mountain, he suggests a retreat from anthropocentrism that is not quite a withdrawal from modern civilization: "We have to un-blind ourselves to human exceptionalism. That's the real challenge. Unless forest-health is our health, we're never going to get beyond appetite as a motivator in the world. The exciting challenge," he says, is to make people "plant-conscious."

IN THEIR ASPIRATIONAL GRANDEUR, ALL THESE TERMS SUGGEST the holistic prospective of a new philosophy, and new ethics, ushered into being by a new world. A raft of popular recent books aims to do the same, their titles so plaintive you could count their spines like rosary beads. Perhaps the baldest entry is Roy Scranton's *Learning to Die in the Anthropocene*. In it, the author, a veteran of the Iraq War, writes, "The greatest challenge we face is a philosophical one: understanding that this civilization is already dead." His subsequent book of essays is *We're Doomed. Now What?*

All these works portend a turn toward the apocalyptic, whether literal, cultural, political, or ethical. But another turn is possible, too, even probable, and perhaps the more tragic for its conspicuous plausibility: that the preponderance of our reflexes in the face of human strife run in the opposite direction, toward acclimatization.

This is the yowling torque muffled by the bland-seeming phrase "climate apathy," which may otherwise feel merely descriptive: that through appeals to nativism, or by the logic of budget realities, or in perverse contortions of "deservedness," by drawing our circles of empathy smaller and smaller, or by simply turning a blind eye when convenient, we will find ways to engineer new indifference. Gazing out at the future from the promontory of the present, with the planet having warmed one degree, the world of two degrees seems nightmarish—and the worlds of three degrees, and four, and five yet more grotesque. But one way we might manage to navigate that path without crumbling collectively in despair is, perversely, to normalize climate suffering at the same pace we accelerate it, as we have so much human pain over centuries, so that we are always coming to terms with what is just ahead of us, decrying what lies beyond that, and forgetting all that we had ever said about the absolute moral unacceptability of the conditions of the world we are passing through in the present tense, and blithely.

IV

The Anthropic Principle

W HAT IF WE'RE WRONG? PERVERSELY, DECADES OF CLI-
mate denial and disinformation have made global warming not merely an ecological crisis but an incredibly high-stakes
wager on the legitimacy and validity of science and the scientific
method itself. It is a bet that science can win only by losing. And in
this test of the climate we have a sample size of just one.

No one wants to see disaster coming, but those who look, do.
Climate science has arrived at this terrifying conclusion not casually, and not with glee, but by systematically ruling out every
alternative explanation for observed warming—even though that
observed warming is more or less precisely what would be expected
given only the rudimentary understanding of the greenhouse effect
advanced by John Tyndall and Eunice Foote in the 1850s, when
America was reaching its first industrial peak. What we are left
with is a set of predictions that can appear falsifiable—about global
temperatures, sea-level rise, and even hurricane frequency and
wildfire volume. But, all told, the question of how bad things will
get is not actually a test of the science; it is a bet on human activity.
How much will we do to stall disaster, and how quickly?

Those are the only questions that matter. There are, it is true,
feedback loops we don't understand and dynamic warming processes scientists haven't yet pinpointed. Yet to the extent we live
today under clouds of uncertainty about climate change, those
clouds are projections not of collective ignorance about the natural
world but blindness about the human one, and can be dispersed
by human action. This is what it means to live beyond the "end of
nature"—that it is human action that will determine the climate of
the future, not systems beyond our control. And it's why, despite
the unmistakable clarity of the predictive science, all of the tentative sketches of climate scenarios that appear in this book are so
oppressively caveated with *possibly*s and *perhaps*es and *conceivably*s.

The emergent portrait of suffering is, I hope, horrifying. It is also, entirely, elective. If we allow global warming to proceed, and to punish us with all the ferocity we have fed it, it will be because we have chosen that punishment—collectively walking down a path of suicide. If we avert it, it will be because we have chosen to walk a different path, and endure.

These are the disconcerting, contradictory lessons of global warming, which counsels both human humility and human grandiosity, each drawn from the same perception of peril. The climate system that gave rise to the human species, and to everything we know of as civilization, is so fragile that it has been brought to the brink of total instability by just one generation of human activity. But that instability is also a measure of the human power that engineered it, almost by accident, and which now must stop the damage, in only as much time. If humans are responsible for the problem, they must be capable of undoing it. We have an idiomatic name for those who hold the fate of the world in their hands, as we do: gods. But for the moment, at least, most of us seem more inclined to run from that responsibility than embrace it—or even admit we see it, though it sits in front of us as plainly as a steering wheel.

Instead, we assign the task to future generations, to dreams of magical technologies, to remote politicians doing a kind of battle with profiteering delay. This is why this book is also studded so oppressively with "we," however imperious it may seem. The fact that climate change is all-enveloping means it targets all of us, and that we must all share in the responsibility so we do not all share in the suffering—at least not all share in so suffocatingly much of it.

We do not know the precise shape such suffering would take, cannot predict with certainty exactly how many acres of forest will burn each year of the next century, releasing into the air centuries of stored carbon; or how many hurricanes will flatten each Caribbean island; or where megadroughts are likely to produce mass famines first; or which will be the first great pandemic to be produced by global warming. But we know enough to see, even now, that the new world we are stepping into will be so alien from our own, it might as well be another planet entirely.

IN 1950, WALKING TO LUNCH AT LOS ALAMOS, THE ITALIAN-born physicist Enrico Fermi, one of the architects of the atom bomb, found himself caught up in a conversation about UFOs with Edward Teller, Emil Konopinski, and Herbert York—so caught up that he drifted off in thought, jumping back in long after everyone else had moved on to ask, "Where is everybody?" The story has now passed into scientific legend, the interjection known as Fermi's paradox: If the universe is so big, then why haven't we encountered any other intelligent life in it?

The answer may be as simple as climate. Nowhere else in the known universe is a single planet as suited as this one to produce life of the kind we know, as Fermi's only children. Global warming makes the proposition seem even more precarious. For the entire historical window in which human life evolved, almost all of the planet has been, climatologically speaking, quite comfortable for us; that is how we managed to get here. But it wasn't always the case even on Earth, where it is no longer comfortable, and only getting less so. No human has ever lived on a planet as hot as this one; it will get hotter. In talking about that near future, several climate scientists I spoke with proposed global warming as a Fermi solution. The natural lifespan of a civilization may only be several thousand years long, and the lifespan of an industrial civilization conceivably only several hundred. In a universe that is many billions of years old, with star systems separated as much by time as by space, civilizations might emerge and develop and then burn themselves up simply too fast to ever find one another.

The Fermi paradox has also been called "the Great Silence"—we bellow out into the universe and hear no echo, and no reply. The iconoclastic economist Robin Hanson calls it "the great filter." Being filtered, the theory goes, are whole civilizations, enclosed by global warming like bugs in a net. "Civilizations rise, but there's an environmental filter that causes them to die off again and disappear fairly quickly," as the charismatic paleontologist Peter Ward—among those responsible for discovering that the planet's mass extinctions were caused by greenhouse gas—told me. "The filtering we've had in the past has been in these mass extinctions." The mass extinction we are now living through has only just begun; so much more dying is coming.

The search for alien life has always been powered by the desire for human importance in a vast, forgetful cosmos: we want to be seen so that we know we exist. What's unusual is that, unlike religion or nationalism or conspiracy theory, the alien fantasy doesn't place humans at the center of a grand story. In fact, it displaces us—in that way it is a sort of Copernican dream. When Copernicus announces the earth revolves around the sun, he briefly feels himself in the spotlight of the universe, but by making the discovery he consigns all of humanity to the relative periphery. This is what my father-in-law, describing what happens to men with the birth of children and then grandchildren, calls "the outer ring theory," and it more or less encapsulates the meaning of any imagined alien encounter: suddenly humans are major players in a drama of almost inconceivable scale, the lasting lesson of which, unfortunately, is that we're total nobodies—or, at best, a lot less unique and important than we thought we were. When the astronauts aboard *Apollo 8* first caught a glimpse of Earth from the tin can carrying them through space—first saw the half-shadowed planet past the surface of the moon—they looked at one another and jokingly asked, about the world that had launched them into orbit, "Is it inhabited?"

In recent years, their telescopes gazing farther, astronomers have discovered legions of planets like our own, many more than were expected a generation ago. This has led to a flurry of activity revising the terms of expectation established by Frank Drake in what is now known as the Drake equation—which builds a prediction about the possibility of extraterrestrial life off assumptions about things like the fraction of planets conceivably able to support life that actually do support life, the fraction of those planets that develop intelligent life, and the fraction of those planets that would emit detectable signs of that intelligence into space.

And there have been many theories, beyond the Great Filter, about why we haven't heard from anybody. There is the "zoo hypothesis," which suggests that aliens are just watching over us and letting us be for now, presumably until we reach their level of sophistication; and something like its inverse—that we haven't heard from aliens because they're the ones sleeping, in a civilization-scale system of extended-sleep pods like the ones we know from science fiction spaceships, waiting while the universe evolves a shape more suitable to their needs. As far back as 1960, the polymath physicist Freeman Dyson proposed that we may be unable to find alien life in our telescopes because advanced civilizations may have literally closed themselves off from the rest of space—encasing whole solar systems in megastructures designed to capture the energy of a central star, a system so efficient that from elsewhere in the universe it would not appear to glow. Climate change suggests another kind of sphere, manufactured not out of technological mastery but first through ignorance, then indolence, then indifference—a civilization enclosing itself in a gaseous suicide, a running car in a scaled garage.

The astrophysicist Adam Frank calls this kind of thinking "the astrobiology of the Anthropocene" in his *Light of the Stars,* which considers climate change, the future of the planet, and our stewardship of it from the perspective of the universe—"thinking like a planet," he calls it. "We are not alone. We are not the first," Frank writes in the book's opening pages. "*This*—meaning everything you see around you in our project of civilization—has quite likely happened thousands, millions, or even trillions of times before."

What sounds like a parable from Nietzsche is really just an explication of the meaning of "infinity," and how small and insignificant the concept makes humans and everything we do in the space of such a universe. In an unconventional recent paper with climatologist Gavin Schmidt, Frank went even further, suggesting that there may even have been advanced industrial civilizations of some form in the deep history of this planet, so deep in the past their remnants would have long been reduced to dust below our feet, making them permanently invisible to us. The paper was meant as a thought experiment, pointing out how little we can really know from archaeology and geology, not a serious claim about the history of the planet.

It was also meant to be uplifting. Frank wanted to offer what he believes is the empowering perspective that our "project of civilization" is profoundly fragile, and that we must take extraordinary measures to protect it. Both are true, but nevertheless it can be a bit hard to see things his way. If there have really been trillions of other civilizations like this one, somewhere out there in the universe and including possibly a few scattered in the dust of the earth, then—whatever lessons of stewardship we can draw from them—it does not bode well for ours that we don't yet see the trace of a single one that's survived.

That is a lot of despair to hang on "trillions"—in fact a lot to hang on some very speculative math. Which goes even more so for the work of anyone trying to "solve" the Drake equation, as many have. That project, which looks to me less like sorting out the nature of the universe on a chalkboard than like playing games with numbers, working from close-to-arbitrary postulates so confidently that when you see the universe departing from your predictions you choose to believe it is hiding from you some very important information—namely, about all the civilizations that may have died and disappeared—rather than that your suppositions may have been made in error. The fact of dramatic near-term climate change should inspire both humility and grandiosity, but this Drakean approach seems to me somehow to get the lesson both right and backward: supposing that the terms of your thought experiment should govern the meaning of the universe, yet unable to imagine that humans might make for themselves an exceptional fate within it.

Fatalism has a strong pull in a time of ecological crisis, but even so it is a curious quirk of the Anthropocene that the transformation of the planet by anthropogenic climate change has produced such fervor for Fermi's paradox and so little for its philosophical counterpoint, the anthropic principle. That principle takes the human anomaly not as a puzzle to explain away but as the centerpiece of a grandly narcissistic view of the cosmos. It's the closest thing string-theory physics can bring us to empowering self-centeredness: that however unlikely it may seem that intelligent civilization arose in an infinity of lifeless gas, and however lonely we appear to be in the universe, in fact something like the world we live on and the one we've built are a sort of logical inevitability, given that we are asking these questions at all—because only a universe compatible with our sort of conscious life would produce anything capable of contemplating it like this.

This is a Möbius strip of a parable, a sort of gimmicky tautology rather than a truth claim based strictly in observed data. And yet, I think, it is much more helpful than Fermi or Drake in thinking about climate change and the existential challenge of solving it in just the few decades ahead. There is one civilization we know of, and it is still around, and kicking—for now, at least. Why should we be suspicious of our exceptionality, or choose to understand it only by assuming an imminent demise? Why not choose to feel empowered by it?

A SENSE OF COSMIC SPECIALNESS IS NO GUARANTEE OF GOOD stewardship. But it does helpfully focus attention on what we are doing to this special planet. You don't need to invoke some imagined law of the universe—that all civilizations are kamikaze ones—to explain the wreckage. You need look only at the choices we have made, collectively; and, collectively, we are at present choosing to wreck it.

Will we stop? "Thinking like a planet" is so alien to the perspectives of modern life—so far from thinking like a neoliberal subject in a ruthless competitive system—that the phrase sounds at first lifted from kindergarten. But reasoning from first principles is reasonable when it comes to climate; in fact, it is necessary, as we only have a first shot to engineer a solution. This goes beyond thinking like a planet, because the planet will survive, however terribly we poison it; it is thinking like a people, one people, whose fate is shared by all.

The path we are on as a planet should terrify anyone living on it, but, thinking like one people, all the relevant inputs are within our control, and there is no mysticism required to interpret or command the fate of the earth. Only an acceptance of responsibility. When Robert Oppenheimer, the actual head of Los Alamos, later reflected on the meaning of the bomb, he famously said he was reminded, in the flash of the first successful nuclear test, of a passage from the Bhagavad-Gita: "Now I am become death, the destroyer of worlds." But the interview was years later, when Oppenheimer had become the pacifist conscience of America's nuclear age—for which, naturally, he had his security clearance revoked. According to his brother, Frank, who was also there when Oppenheimer watched the detonation of the device nicknamed "the gadget," he said only, "It worked."

THE THREAT FROM CLIMATE CHANGE IS MORE TOTAL THAN from the bomb. It is also more pervasive. In a 2018 paper, forty-two scientists from around the world warned that, in a business-as-usual scenario, no ecosystem on Earth was safe, with transformation "ubiquitous and dramatic," exceeding in just one or two centuries the amount of change that unfolded in the most dramatic periods of transformation in the earth's history over tens of thousands of years. Half of the Great Barrier Reef has already died, methane is leaking from Arctic permafrost that may never freeze again, and the high-end estimates for what warming will mean for cereal crops suggest that just four degrees of warming could reduce yields by 50 percent. If this strikes you as tragic, which it should, consider that we have all the tools we need, today, to stop it all: a carbon tax and the political apparatus to aggressively phase out dirty energy; a new approach to agricultural practices and a shift away from beef and dairy in the global diet; and public investment in green energy and carbon capture.

That the solutions are obvious, and available, does not mean the problem is anything but overwhelming. It is not a subject that can sustain only one narrative, one perspective, one metaphor, one mood. This will become only more so in the coming decades, as the signature of global warming appears on more and more disasters, political horrors, and humanitarian crises. There will be those, as there are now, who rage against fossil capitalists and their political enablers; and others, as there are now, who lament human short-sightedness and decry the consumer excesses of contemporary life. There will be those, as there are now, who fight as unrelenting activists, with approaches as diverse as federal lawsuits and aggressive legislation and small-scale protests of new pipelines; nonviolent resistance; and civil-rights crusades. And there will be those, as there are now, who see the cascading suffering and fall back into an inconsolable despair. There will be those, as there are now, who insist that there is only one way to respond to the unfolding ecological catastrophe—one productive way, one responsible way.

Presumably, it won't be only one way. Even before the age of climate change, the literature of conservation furnished many metaphors to choose from. James Lovelock gave us the Gaia hypothesis, which conjured an image of the world as a single, evolving quasi-biological entity. Buckminster Fuller popularized "spaceship earth," which presents the planet as a kind of desperate life raft in what Archibald MacLeish called "the enormous, empty night"; today, the phrase suggests a vivid picture of a world spinning through the solar system barnacled with enough carbon capture plants to actually stall out warming, or even reverse it, restoring as if by magic the breathability of the air between the machines. The *Voyager 1* space probe gave us the "Pale Blue Dot"—the inescapable smallness, and fragility, of the entire experiment we're engaged in, together, whether we like it or not. Personally, I think that climate change itself offers the most invigorating picture, in that even its cruelty flatters our sense of power, and in so doing calls the world, as one, to action. At least I hope it does. But that is another meaning of the climate kaleidoscope. You can choose your metaphor. You can't choose the planet, which is the only one any of us will ever call home.

Acknowledgments

If this book is worth anything, it is worth that because of the work of the scientists who first theorized, then documented the warming of the planet, and then began examining and explicating what that warming might mean for the rest of us living on it. That line of debt runs from Eunice Foote and John Tyndall in the nineteenth century to Roger Revelle and Charles David Keeling in the twentieth and on to all of those hundreds of scientists whose labor appears in the endnotes of this book (and of course many hundreds of unmentioned others very hard at work). However much progress we manage against the assaults of climate change in the coming decades, it is thanks to them.

I am personally indebted to those scientists, climate writers, and activists who were especially generous to me, over the last several years, with their time and insights—helping me understand their own research and pointing me to the findings of others, indulging my requests for rambling interviews or discussing the state of the planet with me in other public settings, corresponding with me over time, and, in many cases, reviewing my writing, including portions of the text of this book, before publication. They are Richard Alley, David Archer, Craig Baker-Austin, David Battisti, Peter Brannen, Wallace Smith Broecker, Marshall Burke, Ethan D. Coffel, Aiguo Dai, Peter Gleick, Jeff Goodell, Al Gore, James Hansen, Katherine Hayhoe, Geoffrey Heal, Solomon Hsiang, Matthew Huber, Nancy Knowlton, Robert Kopp, Lee Kump, Irakli Loladze, Charles Mann, Geoff Mann, Michael Mann, Kate Marvel, Bill McKibben, Michael Oppenheimer, Naomi Oreskes, Andrew Revkin, Joseph Romm, Lynn Scarlett, Steven Sherwood, Joel Wainwright, Peter D. Ward, and Elizabeth Wolkovich.

When I first wrote about climate change in 2017, I relied also on the critical research help of Julia Mead and Ted Hart. I am grateful, too, to all the responses to that story that were published elsewhere—especially those by Genevieve Guenther, Eric Holthaus, Farhad Manjoo, Susan Mathews, Jason Mark, Robinson Meyer, Chris Mooney, and David Roberts. That includes all the scientists who reviewed my work for the website Climate Feedback, working through my story line by line. In preparing this manuscript for publication, Chelsea Leu reviewed it even more closely, and incisively, and I cannot thank her enough for that.

This book would not have come to be without the vision, guidance, wisdom, and forbearance of Tina Bennett, to whom I now owe a lifetime of thanks. And it would not have become an actual book without the acuity, brilliance, and faith of Tim Duggan, and the enormously helpful work of Molly Stern, Dyana Messina, Julia Bradshaw, William Wolfslau, Aubrey Martinson, Julie Cepler, Rachel Aldrich, Craig Adams, Phil Leung, and Andrea Lau, as well as Helen Conford at Penguin in London.

I would not be writing this book were it not for Central Park East, and especially Pam Cushing, my second mother. I'm grateful to everyone I work with at *New York* magazine for all of their encouragement and support along the way. This goes especially for my bosses Jared Hohlt, Adam Moss, and Pam Wasserstein, and

David Haskell, my editor and friend and co-conspirator. Other friends and co-conspirators also helped refine and reconceive what it was I was trying to do in this book, and to all of them I am so thankful, too: Isaac Chotiner, Kerry Howley, Hua Hsu, Christian Lorentzen, Noreen Malone, Chris Parris-Lamb, Willa Paskin, Max Read, and Kevin Roose. For a million unenumerable things, I'd also like to thank Jerry Saltz and Will Leitch, Lisa Miller and Vanessa Grigoriadis, Mike Marino and Andy Roth and Ryan Langer, James Darnton and Andrew Smeall and Scarlet Kim and Ann Fabian, Casey Schwartz and Marie Brenner, Nick Zimmerman and Dan Weber and Whitney Schubert and Joey Frank, Justin Pattner and Daniel Brand, Caitlin Roper, Ann Clarke and Alexis Swerdloff, Stella Bugbee, Meghan O'Rourke, Robert Asahina, Philip Gourevitch, Lorin Stein, and Michael Grunwald.

My best reader, as always, is my brother, Ben; without his footsteps to follow, who knows where I'd be. I've been inspired, too, in countless ways, by Harry and Roseann, Jenn and Matt and Heather, and above all by my mother and father, only one of whom is here to read this book but to both of whom I owe it, and everything else.

The last and biggest thanks belong to Risa, my love, and to Rocca, my other love—for the last year, the last twenty, and the fifty or more to come. Let's hope they're cool ones.

Notes

All science is speculative to some degree, subject to some future reconsideration or revision. But just how speculative varies from science to science, from specialty to specialty, indeed from study to study.

Within climate change research, both the fact of global warming (about 1.1 degrees Celsius since humans first began burning fossil fuels) and its mechanism (the greenhouse gases produced by that burning trap heat radiating upward into the planet's atmosphere) are, at this point, established beyond any shadow of a doubt. Exactly how that warming will play out, over the next decades and then the next centuries, is less certain, both because we don't know how quickly humans will drop their addiction to fossil fuels, and because we don't know precisely how the climate system will recalibrate in response to human perturbation. But the notes that follow are, I hope, a road map to the state of that science, in addition to being a bibliography for this book.

I.   Cascades

3   **five mass extinctions:** Those are the end-Ordovician, the Late Devonian, the end-Permian, the end-Triassic, and the end-Cretaceous. A very good recent popular account of each can be found in Peter Brannen, *The Ends of the World* (New York: HarperCollins, 2017).

3   **86 percent of all species:** These figures are all estimates, and different studies often come to different conclusions. Some accounts of the end-Permian extinction, for instance, suggest the extinction level is as low as 90 percent, while others are as high as 97 percent. These particular figures come from the *Cosmos* primer "The Five Big Mass Extinctions," https://cosmosmagazine .com/palaeontology/big-five-extinctions.

3   **all but the one:** Brannen, *Ends of the World*.

3   **began when carbon warmed:** There is some considerable debate about the precise mix of environmental factors (volcanic eruptions, microbial activity, Arctic methane) that brought about the end-Permian extinction, but for a summary of the theory that volcanic activity warmed the planet and the warming released methane that accelerated that warming, see Uwe Brand et al., "Methane Hydrate: Killer Cause of Earth's Greatest Mass Extinction," *Paleoworld* 25, no. 4 (December 2016): pp. 496–507, https://doi.org/10.1016/ j.palwor.2016.06.002.

4   **at least ten times faster:** "Maximum rates of carbon emissions for both the PETM and the end-Permian are about one billion tons of carbon, and right now we're at ten billion tons of carbon," the Penn State geoscientist Lee Kump, among the world's leading experts on mass extinctions, told me. "The duration of both of those events was much longer than fossil-fuel burning will go on, and so the total amount is lower—but not by a factor of ten. By a factor of two or three."

4   **The rate is one hundred times faster:** Jessica Blunden, Derek S. Arndt, and Gail Hartfield, eds., "State of the Climate in 2017," *Bulletin of the American Meteorological Society* 99, no. 8 (August 2018), Si–S310, https://doi .org/10.1175/2018BAMSStateoftheClimate.1.

4   **at any point in the last 800,000 years:** Rob Moore, "Carbon Dioxide in the Atmosphere Hits Record High Monthly Average," Scripps Institution of Oceanography, May 2, 2018. As Moore puts it: "Prior to the onset of the Industrial Revolution, CO_2 levels had fluctuated over the millennia but had never exceeded 300 ppm at any point in the last 800,000 years," https:// scripps.ucsd.edu/programs/keelingcurve/2018/05/02/carbon-dioxide-in-the -atmosphere-hits-record-high-monthly-average/.

4   **as long as 15 million years:** See, for instance, Aradhna K. Tripati, Christopher D. Roberts, and Robert A. Eagle, "Coupling of CO_2 and Ice Sheet Stability over Major Climate Transitions of the Last 20 Million Years," *Science* 326, no. 5958 (December 2009): pp. 1394–97. "The last time carbon dioxide levels were apparently as high as they are today—and were sustained at those levels—global temperatures were 5 to 10 degrees Fahrenheit higher than they are today," Tripati said in the UCLA press release for the study. "The sea level was approximately 75 to 120 feet higher than today, there was no permanent sea ice cap in the Arctic and very little ice on Antarctica and Greenland."

4   **more than a hundred feet higher:** Ibid.

4   **more than half of the carbon:** Carbon Dioxide Information Analysis Center, Oak Ridge National Laboratory, "Global, Regional, and National Fossil-Fuel CO_2 Emissions" (Oak Ridge, TN, 2017), https://doi.org/10.3334/CDIAC/00001_V2017. Accounts and estimates of historical emissions vary, but according to the Oak Ridge National Laboratory, we have emitted 1578 gigatons of CO_2 from fossil fuels since 1751; since 1989 the total is 820 gigatons.

4   **the figure is about 85 percent:** According to Oak Ridge, the total figure since 1946 is 1376 gigatons, or 87 percent of 1578.

5   **Scientists had understood:** R. Revelle and H. Suess, "Carbon Dioxide Exchange Between Atmosphere and Ocean and the Question of an Increase of Atmospheric CO_2 During the Past Decades," *Tellus* 9 (1957): pp. 18–27.

5   **passing the threshold of carbon concentration:** See, for instance, Nicola Jones, "How the World Passed a Carbon Threshold and Why It Matters," *Yale Environment 360*, January 26, 2017, https://e360.yale.edu/features/how-the-world-passed-a-carbon-threshold-400ppm-and-why-it-matters.

5   **a monthly average of 411:** Scripps Institution of Oceanography, "Another Climate Milestone Falls at Mauna Loa Observatory," June 7, 2018, https://scripps.ucsd.edu/news/another-climate-milestone-falls-mauna-loa-observatory.

6   **more than four degrees Celsius of warming:** IPCC, *Climate Change 2014: Synthesis Report, Summary for Policymakers* (Geneva, 2014), p. 11, www.ipcc.ch/pdf/assessment-report/ar5/syr/AR5_SYR_FINAL_SPM.pdf.

6   **would be rendered uninhabitable:** Gaia Vince, "How to Survive the Coming Century," *New Scientist*, February 25, 2009. Some of this assessment is a bit extreme, but it is incontrovertibly true that warming on that scale will render large parts of those regions brutally inhospitable by any standard we apply today.

7   **a group of Arctic scientists:** Alec Luhn and Elle Hunt, "Besieged Russian Scientists Drive Away Polar Bears," *The Guardian*, September 14, 2016.

7   **killed by anthrax released:** Michaeleen Doucleff, "Anthrax Outbreak in Russia Thought to Be Result of Thawing Permafrost," NPR, August 3, 2016.

7   **one million Syrian refugees:** Phillip Connor, "Most Displaced Syrians Are in the Middle East, and About a Million Are in Europe," Pew Research, January 29, 2018, http://www.pewresearch.org/fact-tank/2018/01/29/where-displaced-syrians-have-resettled.

7   **likely flooding of Bangladesh:** "By 2050, it is estimated that one in every seven people in Bangladesh is likely to be displaced by climate change," Robert Watkins of the United Nations said in a 2015 statement: see Mubashar Hasan, "Bangladesh's Climate Change Migrants," ReliefWeb, November 13, 2015.

7   **140 million by 2050:** World Bank, *Groundswell: Preparing for Internal Climate Migration* (Washington, D.C., 2018), p. xix, https://openknowledge.worldbank.org/handle/10986/29461.

7   **more than a hundred times Europe's Syrian "crisis":** Connor, "Most Displaced Syrians." "Nearly 13 million Syrians are displaced after seven years of conflict in their country," Connor reported.

7   **The U.N. projections are bleaker:** Baher Kamal, "Climate Migrants Might Reach One Billion by 2050," ReliefWeb, August 21, 2017, https://reliefweb .int/report/world/climate-migrants-might-reach-one-billion-2050.

7   **Two hundred million was the entire:** U.S. Census Bureau, "Historical Estimates of World Population," www.census.gov/data/tables/time-series/demo/ international-programs/historical-est-worldpop.html.

7   **"a billion or more vulnerable":** United Nations Convention to Combat Desertification, "Sustainability. Stability. Security," www.unccd.int/ sustainability-stability-security.

8   **Fifteen percent of all human experience:** Eukaryote, "The Funnel of Human Experience," *LessWrong*, October 9, 2018, www.lesswrong.com/posts/SwBE JapZNzWFifLN6/the-funnel-of-human-experience.

9   **another name for that level of warming:** "Marshalls Likens Climate Change Migration to Cultural Genocide," Radio New Zealand, October 6, 2015, www.radionz.co.nz/news/pacific/286139/marshalls-likens-climate-change- migration-to-cultural-genocide.

9   **bell curve of more horrific possibilities:** Technically, this is not a bell curve but a distribution curve, because it has a long tail of negative outcomes, rather than a balanced distribution of optimistic and pessimistic scenarios (that is, there are many more worst-case-like outcomes that are possible than best- case-like outcomes).

11   **about 3.2 degrees of warming:** Perhaps the best reference for all of the various predictive models is the Climate Action Tracker, which calculates that all of the world's existing pledges would likely yield global warming of 3.16 degrees Celsius by 2100.

11   **planet's ice sheets:** Alexander Nauels et al., "Linking Sea Level Rise and Socioeconomic Indicators Under the Shared Socioeconomic Pathways," *Environmental Research Letters* 12, no. 11 (October 2017), https://doi .org/10.1088/1748-9326/aa92b6. In 2017, Nauels and his colleagues suggested that warming of merely 1.9 degrees Celsius could push the ice sheets past a tipping point of collapse.

11   **That would eventually flood:** The total collapse of the ice sheets would raise sea levels by more than two hundred feet, it is estimated, but a much smaller rise would be necessary to flood these cities. Miami sits six feet above sea level, Dhaka thirty-three feet. Shanghai is at thirteen feet, and parts of Hong Kong are as low as zero feet—which is why, in 2015, the *South China Morning Post* reported that four degrees of warming could displace 45 million people in those two cities: Li Ching, "Rising Sea Levels Set to Displace 45 Million People in Hong Kong, Shanghai and Tianjin If Earth Warms 4 Degrees from Climate Change," *South China Morning Post*, November 9, 2015.

11   **several recent studies:** Thorsten Mauritsen and Robert Pincus, "Committed Warming Inferred from Observations," *Nature Climate Change*, July 31, 2017; Adrian E. Raftery et al., "Less than 2°C Warming by 2100 Unlikely," *Nature Climate Change*, July 31, 2017; Hubertus Fischer et al., "Paleoclimate Constraints on the Impact of 2°C Anthropogenic Warming and Beyond," *Nature Geoscience*, June 25, 2018.

12   **"century of hell":** Brady Dennis and Chris Mooney, "Scientists Nearly Double Sea Level Rise Projections for 2100, Because of Antarctica," *The Washington Post*, March 30, 2016.

12   **underestimating the amount of warming:** Alvin Stone, "Global Warming May Be Twice What Climate Models Predict," UNSW Sydney, July 5, 2018, https://newsroom.unsw.edu.au/news/science-tech/global-warming-may-be-twice-what-climate-models-predict.

12   **fire-dominated savanna:** Fischer, "Paleoclimate Constraints on the Impact."

12   **"Hothouse Earth":** Will Steffen et al., "Trajectories of the Earth System in the Anthropocene," *Proceedings of the National Academy of Sciences* (August 14, 2018).

12   **At two degrees, the ice sheets:** Nauels, "Linking Sea Level Rise and Socioeconomic Indicators," https://doi.org/10.1088/1748-9326/aa92b6.

12   **400 million more people:** Robert McSweeney, "The Impacts of Climate Change at 1.5C, 2C and Beyond," *Carbon Brief*, October 4, 2018, https://interactive.carbonbrief.org/impacts-climate-change-one-point-five-degrees-two-degrees.

12   **thirty-two times as many:** Ibid.

13   **9 percent more heat-related deaths:** Ana Maria Vicedo-Cabrera et al., "Temperature-Related Mortality Impacts Under and Beyond Paris Agreement Climate Change Scenario," *Climatic Change* 150, no. 3–4 (October 2018): pp. 391–402, https://doi.org/10.1007/s10584-018-2274-3.

13   **eight million more cases of dengue:** Felipe J. Colon-Gonzalez et al., "Limiting Global-Mean Temperature Increase to 1.5–2 °C Could Reduce the Incidence and Spatial Spread of Dengue Fever in Latin America," *Proceedings of the National Academy of Sciences* 115, no. 24 (June 2018): pp. 6243–48, https://doi.org/10.1073/pnas.1718945115.

13   **The last time that was the case:** As with all work in paleoclimate, estimates on this point vary, but this summary comes from Howard Lee, "What Happened the Last Time It Was as Warm as It's Going to Get at the End of This Century," *Ars Technica*, June 18, 2018.

13   **"hyperobject":** Timothy Morton, *Hyperobjects: Philosophy and Ecology After the End of the World* (Minneapolis: University of Minnesota Press, 2013).

14   **due for about 4.5 degrees:** IPCC, *Climate Change 2014: Synthesis Report*, p. 11, www.ipcc.ch/pdf/assessment-report/ar5/syr/AR5_SYR_FINAL_SPM.pdf.

14   **As Naomi Oreskes has noted:** For instance, in "The Scientific Consensus on Climate Change: How Do We Know We're Not Wrong?" in *Climate Change: What It Means for Us, Our Children, and Our Grandchildren* (Cambridge, MA: MIT Press, 2014).

14   **Just running those models:** Gernot Wagner and Martin L. Weitzman, *Climate Shock: The Economic Consequences of a Hotter Planet* (Princeton, NJ: Princeton University Press, 2015), pp. 53–55.

14   **the Nobel laureate William Nordhaus:** "If productivity growth is high, global temperature in 2100 is 5.3 °C." William Nordhaus, "Projections and Uncertainties About Climate Change in an Area of Minimal Climate Policies" (working paper, National Bureau of Economic Research, 2016).

14   **humans at the equator:** Steven C. Sherwood and Matthew Huber, "An Adaptability Limit to Climate Change Due to Heat Stress," *Proceedings of the National Academy of Sciences* 107, no. 21 (May 2010): pp. 9552–55, https://doi.org/10.1073/pnas.0913352107.

14   **oceans would eventually swell:** Jason Treat et al., "What the World Would Look Like If All the Ice Melted," *National Geographic*, September 2013.

14   **two-thirds of the world's major cities:** This is a common shorthand climate scientists use, expressed by Katharine Hayhoe in Jonah Engel Bromwich, "Where Can You Escape the Harshest Effects of Climate Change?" *The New York Times*, October 20, 2016. "Two-thirds of the world's biggest cities are within a few feet of sea level," Hayhoe says.

14   **hardly any land on the planet:** If, as David Battisti and Rosamond Naylor theorize, every degree of warming costs 10 to 15 percent of grain yields—with higher temperatures cutting into productivity more than lower ones—eight degrees of global warming will almost entirely eliminate the capacity of the world's existing grain regions to produce food.

14   **tropical disease would reach northward:** As Peter Brannen documents in *Ends of the World,* the last time the world was even five degrees warmer, what we now know as the Arctic was, in places, tropical.

15   **climate is actually less sensitive:** Peter M. Cox et al., "Emergent Constraint on Equilibrium Climate Sensitivity from Global Temperature Variability," *Nature* 553 (January 2018): pp. 319–22.

15   **permanent food deficit:** Mark Lynas, *Six Degrees: Our Future on a Hotter Planet* (New York: HarperCollins, 2007). This book is a valuable road map to the future of warming.

15   **"Half-Earth":** Edward O. Wilson, *Half-Earth: Our Planet's Fight for Life* (New York: W. W. Norton, 2016).

16   **three major hurricanes:** Those were Irma, Katia, and Jose.

16   **"500,000-year event":** Tia Ghose, "Hurricane Harvey Caused 500,000-Year Floods in Some Areas," *Live Science,* September 11, 2017, www.livescience.com/60378-hurricane-harvey-once-in-500000-year-flood.html.

17   **third such flood:** Christopher Ingraham, "Houston Is Experiencing Its Third '500-Year' Flood in Three Years. How Is That Possible?" *The Washington Post,* August 29, 2017.

17   **an Atlantic hurricane hit Ireland:** Hurricane Ophelia, that is.

17   **45 million were flooded:** UNICEF, "16 Million Children Affected by Massive Flooding in South Asia, with Millions More at Risk," September 2, 2017, www.unicef.org/press-releases/16-million-children-affected-massive-flooding-south-asia-millions-more-risk.

17   **"thousand-year flood":** Tom Di Liberto, "Torrential Rains Bring Epic Flash Floods in Maryland in Late May 2018," NOAA Climate.gov, May 31, 2018,

www.climate.gov/news-features/event-tracker/torrential-rains-bring-epic
-flash-floods-maryland-late-may-2018.

17   **record heat waves:** Jason Samenow, "Red-Hot Planet: All-Time Heat Records Have Been Set All over the World During the Past Week," *The Washington Post*, July 5, 2018.

17   **fifty-four died from the heat:** Rachel Lau, "Death Toll Rises to 54 as Quebec Heat Wave Ends," *Global News*, July 6, 2018, https://globalnews.ca/
news/4316878/50-people-now-dead-due-to-sweltering-quebec-heat-wave.

17   **one hundred major wildfires:** Jon Herskovitz, "More than 100 Large Wildfires in U.S. as New Blazes Erupt," Reuters, August 11, 2018, www.reuters
.com/article/us-usa-wildfires/more-than-100-large-wildfires-in-u-s-as-new
-blazes-erupt-idUSKBN1KX00B.

17   **4,000 acres in one day:** "Holy Fire Burns 4,000 Acres, Forcing Evacuations in Orange County," Fox 5 San Diego, August 6, 2018, https://fox5
sandiego.com/2018/08/06/fast-moving-wildfire-forces-evacuations-in
-orange-county/.

17   **300-foot eruption of flames:** Kirk Mitchell, "Spring Creek Fire 'Tsunami'
Sweeps over Subdivision, Raising Home Toll to 251," *Denver Post*, July 5,
2018.

18   **1.2 million were evacuated:** Elaine Lies, "Hundreds of Thousands Evacuated
in Japan as 'Historic Rain' Falls; Two Dead," Reuters, July 6, 2018, https://
af.reuters.com/article/commoditiesNews/idAFL4N1U21AH.

18   **the evacuation of 2.45 million:** "Two Killed, 2.45 Million Evacuated as
Super Typhoon Mangkhut Hits Mainland China," *The Times of India*, September 16, 2018, https://timesofindia.indiatimes.com/world/china/super
-typhoon-mangkhut-hits-china-over-2-45-million-people-evacuated/
articleshow/65830611.cms.

18   **turning the port city of Wilmington:** Patricia Sullivan and Katie Zezima,
"Florence Has Made Wilmington, N.C., an Island Cut Off from the Rest of
the World," *The Washington Post*, September 16, 2018.

18   **hog manure and coal ash:** Umair Irfan, "Hog Manure Is Spilling Out of La-
goons Because of Hurricane Florence's Floods," *Vox*, September 21, 2018.

18   **the winds of Florence:** Joel Burgess, "Tornadoes in the Wake of Florence
Twist Through North Carolina," Asheville *Citizen-Times*, September 17,
2018.

18   **Kerala was hit:** Hydrology Directorate, Government of India, *Study Report: Kerala Floods of August 2018* (September 2018), http://cwc.gov.in/main/
downloads/KeralaFloodReport/Rev-0.pdf.

18   **Hawaii's East Island:** Josh Hafner, "Remote Hawaiian Island Vanishes Underwater After Hurricane," *USA Today*, October 24, 2018.

18   **deadliest fire in its history:** Paige St. John et al., "California Fire: What
Started as a Tiny Brush Fire Became the State's Deadliest Wildfire. Here's
How," *Los Angeles Times*, November 18, 2018.

18   **Jerry Brown described:** Ruben Vives, Melissa Etehad, and Jaclyn Cosgrove,
"Southern California Fire Devastation Is 'the New Normal,' Gov. Brown
Says," *Los Angeles Times*, December 10, 2017.

20   **"angry beast":** "Wallace Broecker: How to Calm an Angry Beast," CBC News, November 19, 2008, www.cbc.ca/news/technology/wallace-broecker -how-to-calm-an-angry-beast-1.714719.

21   **the fourth evacuation order:** County of Santa Barbara, California, evacuation orders from 2018.

21   **temporary shacks:** Michael Schwirtz, "Besieged Rohingya Face 'Crisis Within the Crisis': Deadly Floods," *The New York Times*, February 13, 2018.

21   **More than a dozen died:** Phil Helsel, "Body of Mother Found After California Mudslide; Death Toll Rises to 21," NBC News, January 20, 2018, www .nbcnews.com/news/us-news/body-mother-found-after-california-mudslide -death-toll-rises-21-n839546.

22   **1.8 trillion tons of carbon:** NASA Science, "Is Arctic Permafrost the 'Sleeping Giant' of Climate Change?" NASA, June 24, 2013, https://science.nasa. gov/science-news/science-at-nasa/2013/24jun_permafrost.

22   **thirty-four times as powerful:** Environmental Protection Agency, "Greenhouse Gas Emissions: Understanding Global Warming Potentials," www.epa.gov/ghgemissions/understanding-global-warming-potentials.

22   **climate scientists call "feedbacks":** For a good overview, see Lee R. Kump and Michael E. Mann, *Dire Predictions: The Visual Guide to the Findings of the IPCC*, 2nd ed. (New York: DK, 2015).

23   **human-triggered avalanches:** Melanie J. Froude and David N. Petley, "Global Fatal Landslide Occurrence from 2004 to 2016," *Natural Hazards and Earth Systems Sciences* 18 (2018): pp. 2161–81, https://doi.org/10.5194/ nhess-18-2161-2018.

23   **a whole new kind:** Bob Berwyn, "Destructive Flood Risk in U.S. West Could Triple If Climate Change Left Unchecked," *Inside Climate News* (August 6, 2018), https://insideclimatenews.org/news/06082018/global -warming-climate-change-floods-california-oroville-dam-scientists.

24   **500,000 poor Latinos:** Ellen Wulfhorst, "Overlooked U.S. Border Shantytowns Face Threat of Gathering Storms," Reuters, June 11, 2018, https:// af.reuters.com/article/commoditiesNews/idAFL2N1SO2FZ.

24   **countries with lower GDPs:** Andrew D. King and Luke J. Harrington, "The Inequality of Climate Change from 1.5°C to 2°C of Global Warming," *Geophysical Research Letters* 45, no. 10 (May 2018): pp. 5030–33, https://doi .org/10.1029/2018GL078430.

25   **trees may simply turn brown:** Andrea Thompson, "Drought and Climate Change Could Throw Fall Colors Off Schedule," *Scientific American*, November 1, 2016.

25   **coffee plants of Latin America:** Pablo Imbach et al., "Coupling of Pollination Services and Coffee Suitability Under Climate Change," *Proceedings of the National Academy of Sciences* 114, no. 39 (September 2017): pp. 10438–42, https://doi.org/10.1073/pnas.1617940114. The paper was summarized by Yale's *E360* this way: "Latin America could lose up to 90 percent of its coffee-growing land by 2050."

25   **half of the world's vertebrate animals:** WWF, "Living Planet Report 2018," *Aiming Higher* (Gland, Switz.: 2018), p. 18, https://wwf.panda.org/ knowledge_hub/all_publications/living_planet_report_2018.

26   **the flying insect population declined:** Caspar Hallman et al., "More Than 75 Percent Decline over 27 Years in Total Flying Insect Biomass in Protected Areas," *PLOS One* 12, no. 10 (October 2017), https://doi.org/10.1371/journal .pone.0185809.

26   **delicate dance of flowers and their pollinators:** Damian Carrington, "Climate Change Is Disrupting Flower Pollination, Research Shows," *The Guardian*, November 6, 2014.

26   **migration patterns of cod:** Bob Berwyn, "Fish Species Forecast to Migrate Hundreds of Miles Northward as U.S. Waters Warm," *Inside Climate News*, May 16, 2018, https://insideclimatenews.org/news/16052018/fish-species -climate-change-migration-pacific-northwest-alaska-atlantic-gulf-maine -cod-pollock.

26   **hibernation patterns of black bears:** Kendra Pierre-Louis, "As Winter Warms, Bears Can't Sleep, and They're Getting into Trouble," *The New York Times*, May 4, 2018.

26   **whole new class of hybrid species:** Moises Velaquez-Manoff, "Should You Fear the Pizzly Bear?" *The New York Times Magazine*, August 14, 2014.

26   **desertification of the entire Mediterranean:** Joel Guiot and Wolfgang Cramer, "Climate Change: The 2015 Paris Agreement Thresholds and Mediterranean Basin Ecosystems," *Science* 354, no. 6311 (October 2016): pp. 463–68, https://doi.org/10.1126/science.aah5015. According to Guiot and Cramer's calculations, even staying below two degrees of warming would mean much of the region would become, technically at least, desert.

26   **dust from the Sahara:** "Sahara Desert Dust Cloud Blankets Greece in Orange Haze," Sky News, March 26, 2018, https://news.sky.com/story/ sahara-desert-dust-cloud-blankets-greece-in-orange-haze-11305011.

26   **for the Nile to be dramatically drained:** "How Climate Change Might Affect the Nile," *The Economist*, August 3, 2017.

26   **the Rio Sand:** Tom Yulsman, "Drought Turns the Rio Grande into the 'Rio Sand,'" *Discover*, July 15, 2013.

27   **Eight hundred million in South Asia:** Muthukumara Mani et al., "South Asia's Hotspots: Impacts of Temperature and Precipitation Changes on Living Standards," World Bank (Washington, D.C., June 2018), p. xi, https://openknowledge.worldbank.org/bitstream/handle/10986/28723/ 9781464811555.pdf?sequence=5&isAllowed=y.

27   **fossil capitalism:** Andreas Malm, *Fossil Capital: The Rise of Steam Power and the Roots of Global Warming* (London: Verso, 2016).

27   **about one percentage point of GDP:** Solomon Hsiang et al., "Estimating Economic Damage from Climate Change in the United States," *Science* 356, no. 6345 (June 2017): pp. 1362–69, https://doi.org/10.1126/science.aal4369.

27   **$20 trillion richer:** Marshall Burke et al., "Large Potential Reduction in Economic Damages Under UN Mitigation Targets," *Nature* 557 (May 2018): pp. 549–53, https://doi.org/10.1038/s41586-018-0071-9.

27   **$551 trillion in damages:** R. Warren et al., "Risks Associated with Global Warming of 1.5 or 2C," Tyndall Centre for Climate Change Research, May 2018, www.tyndall.ac.uk/sites/default/files/publications/briefing_note _risks_warren_r1-1.pdf.

27   **total worldwide wealth is today:** According to Credit Suisse's *Global Wealth Report 2017*, total global wealth that year was $280 trillion.

27   **has not topped 5 percent globally:** According to the World Bank, the last time was 1976, when global growth was at 5.355 percent. World Bank, "GDP Growth (Annual %)," https://data.worldbank.org/indicator/NY.GDP.MKTP.KD.ZG.

27   **"steady-state economics":** The term was popularized by Herbert Daly, whose anthology *Toward a Steady-State Economy* (San Francisco: W.H. Freeman, 1973) established a contrarian perspective on the history of economic growth that is especially incisive in an age of climate change. ("The economy is a wholly owned subsidiary of the environment, not the reverse.")

28   **150 million more people:** Drew Shindell et al., "Quantified, Localized Health Benefits of Accelerated Carbon Dioxide Emissions Reductions," *Nature Climate Change* 8 (March 2018): pp. 291–95, https://doi.org/10.1038/s41558-018-0108-y.

28   **IPCC raised the stakes:** IPCC, *Global Warming of 1.5°C: An IPCC Special Report on the Impacts of Global Warming of 1.5°C Above Pre-Industrial Levels and Related Global Greenhouse Gas Emission Pathways, in the Context of Strengthening the Global Response to the Threat of Climate Change, Sustainable Development, and Efforts to Eradicate Poverty* (Incheon, Korea, 2018), www.ipcc.ch/report/sr15.

28   **seven million deaths:** This is from the World Health Organization's 2014 assessment, in which air pollution was named as the single biggest health risk in the world: WHO, "Public Health, Environmental and Social Determinants of Health (PHE)," www.who.int/phe/health_topics/outdoorair/databases/en.

31   **whether it's responsible to have children:** For a useful summary of this suddenly pervasive query among Western liberals and a fairly thorough counterargument, see Connor Kilpatrick, "It's Okay to Have Children," *Jacobin*, August 22, 2018.

32   **Paul Hawken has perhaps illustrated:** You can find his comprehensive survey of climate solutions (plant-based diets, green roofs, the education of women) in *Drawdown: The Most Comprehensive Plan Ever Proposed to Reverse Global Warming* (New York: Penguin, 2017).

32   **Fully half of British emissions:** This is probably an overestimate, but it comes from "Less In, More Out," published by the U.K.'s *Green Alliance* in 2018.

32   **two-thirds of American energy:** Anne Stark, "Americans Used More Clean Energy in 2016," Lawrence Livermore National Laboratory, April 10, 2017, www.llnl.gov/news/americans-used-more-clean-energy-2016.

33   **$5 trillion each year:** David Coady et al., "How Large Are Global Fossil Fuel Subsidies?" *World Development* 91 (March 2017): pp. 11–27, https://doi.org/10.1016/j.worlddev.2016.10.004.

33   **cost the world $26 trillion:** The New Climate Economy, "Unlocking the Inclusive Growth Story of the 21st Century: Accelerating Climate Action in Urgent Times" (Washington, D.C.: Global Commission on the Economy and Climate, September 2018), p. 8, https://newclimateeconomy.report/2018.

33   **Americans waste a quarter of their food:** Zach Conrad et al., "Relationship Between Food Waste, Diet Quality, and Environmental Sustainability," *PLOS One* 13, no. 4 (April 2018), https://doi.org/10.1371/journal.pone.0195405.

33   **mining it consumes more electricity:** Eric Holthaus, "Bitcoin's Energy Use Got Studied, and You Libertarian Nerds Look Even Worse than Usual," *Grist*, May 17, 2018, https://grist.org/article/bitcoins-energy-use-got-studied-and-you-libertarian-nerds-look-even-worse-than-usual. See also Alex de Vries, "Bitcoin's Growing Energy Problem," *Cell* 2, no. 5 (May 2018): pp. 801–5, https://doi.org/10.1016/j.joule.2018.04.016.

33   **Seventy percent of the energy:** Nicola Jones, "Waste Heat: Innovators Turn to an Overlooked Renewable Resource," *Yale Environment 360*, May 29, 2018. "Today, in the United States, most fossil fuel–burning power plants are about 33 percent efficient," Jones writes, "while combined heat and power (CHP) plants are typically 60 to 80 percent efficient."

33   **U.S. carbon emissions:** The World Bank estimated the 2014 U.S. carbon emissions per capita at 16.49 metric tons per year; the average citizen of the E.U., that year, was responsible for just 6.379 (so the savings would actually be considerably more than 50 percent). World Bank, "CO_2 Emissions (Metric Tons per Capita)," https://data.worldbank.org/indicator/EN.ATM.CO2E.PC.

33   **global emissions would fall by a third:** The richest 10 percent of the world are responsible for about half of all emissions, Oxfam calculated in its "Extreme Carbon Inequality" report of December 2015, available at www.oxfam.org/sites/www.oxfam.org/files/file_attachments/mb-extreme-carbon-inequality-021215-en.pdf. The average carbon footprint for someone in the global 1 percent, the study found, was 175 times that of someone in the world's poorest 10 percent.

35   **We have already left behind:** Perhaps the most vivid illustration of this is the xkcd web comic "A Timeline of Earth's Average Temperature," September 12, 2016, www.xkcd.com/1732.

II.   Elements of Chaos

Heat Death

39   **At seven degrees of warming:** Steven C. Sherwood and Matthew Huber, "An Adaptability Limit to Climate Change Due to Heat Stress," *Proceedings of the National Academy of Sciences* 107, no. 21 (May 2010): pp. 9552–55, https://doi.org/10.1073/pnas.0913352107.

39   **after a few hours:** Ibid. According to Sherwood and Huber, "Periods of net heat storage can be endured, though only for a few hours, and with ample time needed for recovery."

39   **eleven or twelve degrees Celsius:** Ibid. "With 11–12°C warming, such regions would spread to encompass the majority of the human population as currently distributed," Sherwood and Huber write. "Eventual warmings of 12°C are possible from fossil fuel burning."

39   **at just five degrees:** Mark Lynas, *Six Degrees: Our Future on a Hotter Planet* (Washington, D.C.: National Geographic Society, 2008), p. 196.

39   **summer labor of any kind:** John P. Dunne et al., "Reductions in Labour Capacity from Heat Stress Under Climate Warming," *Nature Climate Change* 3 (February 2013): pp. 563–66, https://doi.org/10.1038/NCLIMATE1827.

40   **New York City would be hotter:** Joseph Romm, *Climate Change: What Everyone Needs to Know* (New York: Oxford University Press, 2016), p. 138.

40   **median projection of over four degrees:** IPCC, *Climate Change 2014: Synthesis Report*, Summary for Policymakers (Geneva, 2014), p. 11, www.ipcc.ch/pdf/assessment-report/ar5/syr/AR5_SYR_FINAL_SPM.pdf.

40   **fiftyfold increase:** Romm, *Climate Change*, p. 41.

40   **five warmest summers in Europe:** World Bank, *Turn Down the Heat: Why a 4°C Warmer World Must Be Avoided* (Washington, D.C., November 2012), p. 13, http://documents.worldbank.org/curated/en/865571468149107611/pdf/NonAsciiFileName0.pdf.

40   **simply working outdoors:** IPCC, *Climate Change 2014*, p. 15, www.ipcc.ch/pdf/assessment-report/ar5/syr/AR5_SYR_FINAL_SPM.pdf. "By 2100 for RCP8.5, the combination of high temperature and humidity in some areas for parts of the year is expected to compromise common human activities, including growing food and working outdoors."

40   **cities like Karachi and Kolkata:** Tom K. R. Matthews, et al., "Communicating the Deadly Consequences of Global Warming for Human Heat Stress," *Proceedings of the National Academy of Sciences* 114, no. 15 (April 2017): pp. 3861–66, https://doi.org/10.1073/pnas.1617526114. The authors write, of the 2015 summer, "The extraordinary heat had deadly consequences, with over 3,400 fatalities reported across India and Pakistan alone."

40   **European heat wave of 2003:** World Bank, *Turn Down the Heat*, p. 37, http://documents.worldbank.org/curated/en/865571468149107611/pdf/NonAsciiFileName0.pdf.

41   **worst weather events in Continental history:** William Langewiesche, "How Extreme Heat Could Leave Swaths of the Planet Uninhabitable," *Vanity Fair*, August 2017.

41   **a research team led by Ethan Coffel:** Ethan Coffel et al., "Temperature and Humidity Based on Projections of a Rapid Rise in Global Heat Stress Exposure During the 21st Century," *Environmental Research Letters* 13 (December 2017), https://doi.org/10.1088/1748-9326/aaa00e.

41   **the World Bank has estimated:** World Bank, *Turn Down the Heat*, p. 38, http://documents.worldbank.org/curated/en/865571468149107611/pdf/NonAsciiFileName0.pdf.

41   **Indian summer killed 2,500:** IFRC, "India: Heat Wave—Information Bulletin No. 01," June 11, 1998, www.ifrc.org/docs/appeals/rpts98/in002.pdf.

41   **In 2010, 55,000 died:** In Moscow, there were 10,000 ambulance calls each day, and many doctors believed that the official death counts understated the true toll.

41   **according to *The Wall Street Journal*:** Craig Nelson and Ghassan Adan, "Iraqis Boil as Power-Grid Failings Exacerbate Heat Wave," *The Wall Street Journal*, August 11, 2016.

42   **700,000 barrels of oil:** Ayhan Demirbas et al., "The Cost Analysis of Electric Power Generation in Saudi Arabia," *Energy Sources, Part B* 12, no. 6 (March 2017): pp. 591–96, https://doi.org/10.1080/15567249.2016.1248874.

42   **10 percent of global electricity:** International Energy Agency, *The Future of Cooling: Opportunities for Energy-Efficient Air Conditioning* (Paris, 2018), p. 24, www.iea.org/publications/freepublications/publication/The_Future _of_Cooling.pdf.

42   **triple, or perhaps quadruple:** Ibid., p. 3.

42   **700 million AC units:** Nihar Shah et al., "Benefits of Leapfrogging to Superefficiency and Low Global Warming Potential Refrigerants in Room Air Conditioning," Lawrence Berkeley National Laboratory (October 2015), p. 18, http://eta-publications.lbl.gov/sites/default/files/lbnl-1003671.pdf.

42   **more than nine billion cooling appliances:** University of Birmingham, *A Cool World: Defining the Energy Conundrum of Cooling for All* (Birmingham, 2018), p. 3, www.birmingham.ac.uk/Documents/college-eps/energy/ Publications/2018-clean-cold-report.pdf.

42   **hajj will become physically impossible:** Jeremy S. Pal and Elfatih A. B. Eltahir, "Future Temperature in Southwest Asia Projected to Exceed a Threshold for Human Adaptability," *Nature Climate Change* 6 (2016), pp. 197–200, www.nature.com/articles/nclimate2833.

42   **sugarcane region of El Salvador:** Oriana Ramirez-Rubio et al., "An Epidemic of Chronic Kidney Disease in Central America: An Overview," *Journal of Epidemiology and Community Health* 67, no. 1 (September 2012): pp. 1–3, http://dx.doi.org/10.1136/jech-2012-201141.

44   **grew by 1.4 percent:** International Energy Agency, *Global Energy and CO_2 Status Report, 2017* (Paris, March 2018), p. 1, www.iea.org/publications/ freepublications/publication/GECO2017.pdf.

44   **"in range":** See the Climate Action Tracker.

45   **emissions grew by 4 percent:** Zach Boren and Harri Lammi, "Dramatic Surge in China Carbon Emissions Signals Climate Danger," *Unearthed*, May 30, 2018, https://unearthed.greenpeace.org/2018/05/30/ china-co2-carbon-climate-emissions-rise-in-2018.

45   **coal power has nearly doubled:** Simon Evans and Rosamund Pearce, "Mapped: The World's Coal Power Plants," *Carbon Brief*, June 5, 2018, www.carbon brief.org/mapped-worlds-coal-power-plants. Evans and Pearce estimate 1.061 million megawatts of coal power in 2000 and 1.996 million in 2017.

45   **the Chinese example:** Yann Robiou du Pont and Malte Meinshausen, "Warming Assessment of the Bottom-Up Paris Agreement Emissions Pledges," *Nature Communications*, November 2018.

45   **"limited realistic potential":** European Academies' Science Advisory Council, *Negative Emission Technologies: What Role in Meeting Paris Agreement*

Targets? (Halle, Ger., February 2018), p. 1, https://easac.eu/fileadmin/PDF_s/ reports_statements/Negative_Carbon/EASAC_Report_on_Negative _Emission_Technologies.pdf.

45   **"magical thinking":** "Why Current Negative-Emissions Strategies Remain 'Magical Thinking,'" *Nature*, February 21, 2018, www.nature.com/articles/ d41586-018-02184-x.

46   **full-scale carbon capture plants:** Andy Skuce, "'We'd Have to Finish One New Facility Every Working Day for the Next 70 Years'—Why Carbon Capture Is No Panacea," *Bulletin of the Atomic Scientists,* October 4, 2016, https://thebulletin.org/2016/10/wed-have-to-finish-one-new-facility-every -working-day-for-the-next-70-years-why-carbon-capture-is-no-panacea.

46   **eighteen of them:** Global CCS Institute, "Large-Scale CCS Facilities," www.globalccsinstitute.com/projects/large-scale-ccs-projects.

46   **Asphalt and concrete:** Linda Poon, "Street Grids May Make Cities Hot-ter," *CityLab*, April 27, 2018, www.citylab.com/environment/2018/04/ street-grids-may-make-cities-hotter/558845.

47   **22 degrees Fahrenheit:** Environmental Protection Agency, "Heat Island Ef-fect," www.epa.gov/heat-islands.

47   **Chicago heat wave of 1995:** Eric Klinenberg, *Heat Wave: A Social Autopsy of Disaster in Chicago* (Chicago: University of Chicago Press, 2002).

47   **two-thirds of the global population:** "Around 2.5 Billion More People Will Be Living in Cities by 2050, Projects New U.N. Report," United Nations Department of Economic and Social Affairs, May 16, 2018, www.un.org/ development/desa/en/news/population/2018-world-urbanization-prospects .html.

48   **that list could grow to 970:** Urban Climate Change Research Network, *The Future We Don't Want: How Climate Change Could Impact the World's Great-est Cities* (New York, February 2018), p. 6, https://c40-production-images.s3 .amazonaws.com/other_uploads/images/1789_Future_We_Don't_Want_ Report_1.4_hi-res_120618.original.pdf.

48   **70,000 workers:** Public Citizen, "Extreme Heat and Unprotected Work-ers: Public Citizen Petitions OSHA to Protect the Millions of Workers Who Labor in Dangerous Temperatures" (Washington, D.C.: July 17, 2018), p. 25, www.citizen.org/sites/default/files/extreme_heat_and_unprotected _workers.pdf.

48   **255,000 are expected:** World Health Organization, "Quantitative Risk As-sessment of the Effects of Climate Change on Selected Causes of Death, 2030s and 2050s" (Geneva, 2014), p. 21, http://apps.who.int/iris/bitstream/ handle/10665/134014/9789241507691_eng.pdf?sequence=1&isAllowed=y.

48   **a third of the world's population:** Camilo Mora et al., "Global Risk of Deadly Heat," *Nature Climate Change* 7 (June 2017): pp. 501–6, https://doi .org/10.1038/nclimate3322.

48   **heat death is among:** Langewiesche, "How Extreme Heat Could Leave Swaths."

Hunger

49   **the basic rule of thumb:** David S. Battisti and Rosamond L. Naylor, "Historical Warnings of Future Food Insecurity with Unprecedented Seasonal Heat," *Science* 323, no. 5911 (January 2009): pp. 240–44.

49   **Some estimates run higher:** "The temperature-crop relationship is nonlinear," Battisti says. "Yields drop off faster for each one degree Celsius temperature increase—so yes, all else being the same, yields would drop off much more than 50 percent."

49   **eight pounds of grain to produce:** Lloyd Alter, "Energy Required to Produce a Pound of Food," *Treehugger*, 2010. As Battisti put it in an interview, "Usually this is quoted as 'it takes 8 to 10 kg of grain to produce 1 kg of beef.'"

49   **Globally, grain accounts:** Ed Yong, "The Very Hot, Very Hungry Caterpillar," *The Atlantic*, August 30, 2018.

49   **two-thirds of all human calories:** Chuang Zhao et al., "Temperature Increase Reduces Global Yields of Major Crops in Four Independent Estimates," *Proceedings of the National Academy of Sciences* 114, no. 35 (August 2017): pp. 9326–31, https://doi.org/10.1073/pnas.1701762114.

49   **the United Nations estimates:** Food and Agriculture Organization, "How to Feed the World in 2050" (Rome, October 2009), p. 2, www.fao.org/fileadmin/templates/wsfs/docs/expert_paper/How_to_Feed_the_World_in_2050.pdf.

50   **the tropics are already too hot:** "In the tropics, the temperature already exceeds the optimate temperature for major grains," Battisti told me. "Any additional increase in temperature will further reduce yield, even under otherwise optimal conditions."

50   **at least a fifth of its productivity:** Michelle Tigchelaar et al., "Future Warming Increases Probability of Globally Synchronized Maize Production Shocks," *Proceedings of the National Academy of Sciences* 115, no. 26 (June 2018): pp. 6644–49, https://doi.org/10.1073/pnas.1718031115.

50   **thicker leaves are worse:** Marlies Kovenock and Abigail L. S. Swann, "Leaf Trait Acclimation Amplifies Simulated Climate Warming in Response to Elevated Carbon Dioxide," *Global Biogeochemical Cycles* 32 (October 2018), https://doi.org/10.1029/2018GB005883.

51   **75 billion tons of soil:** Stacey Noel et al., "Report for Policy and Decision Makers: Reaping Economic and Environmental Benefits from Sustainable Land Management," Economics of Land Development Initiative (Bonn, Ger., September 2015), p. 10, www.eld-initiative.org/fileadmin/pdf/ELD-pm-report_05_web_300dpi.pdf.

51   **the rate of erosion is ten times:** Susan S. Lang, "'Slow, Insidious' Soil Erosion Threatens Human Health and Welfare as Well as the Environment, Cornell Study Asserts," *Cornell Chronicle*, March 20, 2006, http://news.cornell.edu/stories/2006/03/slow-insidious-soil-erosion-threatens-human-health-and-welfare.

51   **thirty to forty times as fast:** Ibid.

51   **lacking credit to make the necessary:** Richard Hornbeck, "The Enduring Impact of the American Dust Bowl: Short- and Long-Run Adjustments to Environmental Catastrophe," *American Economic Review* 102, no. 4 (June 2012): pp. 1477–507, http://doi.org/10.1257/aer.102.4.1477.

51   **John Wesley Powell:** Richard Seager et al., "Whither the 100th Meridian? The Once and Future Physical and Human Geography of America's Arid-Humid Divide. Part 1: The Story So Far," *Earth Interactions* 22, no. 5 (March 2018), https://doi.org/10.1175/EI-D-17-0011.1. You can read further by finding Powell's own text, "Report on the Lands of the Arid Region of the United States, with a More Detailed Account of the Lands of Utah. With Maps" (Washington, D.C.: Government Printing Office, 1879), https://pubs.usgs.gov/unnumbered/70039240/report.pdf.

51   **less farmable land:** Seager, "Whither the 100th Meridian?" https://doi.org/10.1175/EI-D-17-0011.1.

51   **separating the Sahara desert:** Lamont-Doherty Earth Observatory, "The 100th Meridian, Where the Great Plains Begins, May Be Shifting," April 11, 2018, www.ldeo.columbia.edu/news-events/100th-meridian-where-great-plains-begin-may-be-shifting.

52   **That desert has expanded:** Natalie Thomas and Sumant Nigam, "Twentieth-Century Climate Change over Africa: Seasonal Hydroclimate Trends and Sahara," *Journal of Climate* 31, no. 22 (2018).

52   **dropped from more than 30 percent:** Food and Agriculture Organization, "The State of Food Insecurity in the World: Addressing Food Insecurity in Protracted Crises" (Rome, 2010), p. 9, www.fao.org/docrep/013/i1683e/i1683e.pdf.

53   **Born to Iowa family farmers:** Charles C. Mann, *The Wizard and the Prophet: Two Remarkable Scientists and Their Dueling Visions to Shape Tomorrow's World* (New York: Knopf, 2018).

54   **increase global greenhouse-gas emissions:** Zhaohai Bai et al., "Global Environmental Costs of China's Thirst for Milk," *Global Change Biology* 24, no. 5 (May 2018): pp. 2198–211, https://doi.org/10.1111/gcb.14047.

54   **food production accounts for about a third:** Natasha Gilbert, "One-Third of Our Greenhouse Gas Emissions Come from Agriculture," *Nature*, October 31, 2012, www.nature.com/news/one-third-of-our-greenhouse-gas-emissions-come-from-agriculture-1.11708.

54   **Greenpeace has estimated:** Greenpeace International, "Greenpeace Calls for Decrease in Meat and Dairy Production and Consumption for a Healthier Planet" (press release), March 5, 2018, www.greenpeace.org/international/press-release/15111/greenpeace-calls-for-decrease-in-meat-and-dairy-production-and-consumption-for-a-healthier-planet.

54   **"the Malthusian tragic:"** Kris Bartkus, "W. G. Sebald and the Malthusian Tragic," *The Millions*, March 28, 2018.

55   **At 2 degrees of warming:** Mark Lynas, *Six Degrees: Our Future on a Hotter Planet* (Washington, D.C.: National Geographic Society, 2008), p. 84.

55   **"two globe-girdling belts of perennial drought":** Ibid.

55   **By 2080, without dramatic reductions:** Benjamin I. Cook et al., "Global Warming and 21st Century Drying," *Climate Dynamics* 43, no. 9–10 (March 2014): pp. 2607–27, https://doi.org/10.1007/s00382-014-2075-y.

56   **The same will be true in Iraq and Syria:** Joseph Romm, *Climate Change: What Everyone Needs to Know* (New York: Oxford University Press, 2016), p. 101.

56   **all the rivers east of the Sierra Nevada:** Ibid., p. 102.

56   **100 million hungry:** Food and Agriculture Organization, "The State of Food Security and Nutrition in the World: Building Climate Resilience for Food Security and Nutrition" (Rome, 2018), p. 57, www.fao.org/3/I9553EN/i9553en.pdf.

56   **The spring of 2017 brought:** "Fighting Famine in Nigeria, Somalia, South Sudan and Yemen," ReliefWeb, 2017, https://reliefweb.int/topics/fighting-famine-nigeria-somalia-south-sudan-and-yemen.

56   **truly customized farming strategies:** Zhenling Cui et al, "Pursuing Sustainable Productivity with Millions of Smallholder Farmers," *Nature*, March 7, 2018.

56   **"soil-free startup":** Madeleine Cuff, "Green Growth: British Soil-Free Farming Startup Prepares for First Harvest," *Business Green*, May 1, 2018.

57   **"We are witnessing the greatest injection":** Helena Bottemiller Evich, "The Great Nutrient Collapse," *Politico*, September 13, 2017.

57   **has declined by as much as one-third:** Donald R. Davis et al., "Changes in USDA Food Composition Data for 43 Garden Crops, 1950 to 1999," *Journal of the American College of Nutrition* 23, no. 6 (2004): pp. 669–82.

57   **the protein content of bee pollen:** Lewis H. Ziska et al., "Rising Atmospheric CO_2 Is Reducing the Protein Concentration of a Floral Pollen Source Essential for North American Bees," *Proceedings of the Royal Society B* 283, no. 1828 (April 2016), http://dx.doi.org/10.1098/rspb.2016.0414.

57   **by 2050 as many as 150 million:** Danielle E. Medek et al., "Estimated Effects of Future Atmospheric CO_2 Concentrations on Protein Intake and the Risk of Protein Deficiency by Country and Region," *Environmental Health Perspectives* 125, no. 8 (August 2017), https://doi.org/10.1289/EHP41.

57   **138 million could suffer:** Samuel S. Myers et al., "Effect of Increased Concentrations of Atmospheric Carbon Dioxide on the Global Threat of Zinc Deficiency: A Modelling Study," *The Lancet* 3, no. 10 (October 2015): PE639–E645, https://doi.org/10.1016/S2214-109X(15)00093-5.

57   **1.4 billion could face a dramatic decline:** M. R. Smith et al., "Potential Rise in Iron Deficiency Due to Future Anthropogenic Carbon Dioxide Emissions," *GeoHealth* 1 (August 2017): pp. 248–57, https://doi.org/10.1002/2016GH000018.

57   **eighteen different strains of rice:** Chunwu Zhu et al., "Carbon Dioxide (CO_2) Levels This Century Will Alter the Protein, Micronutrients, and Vitamin Content of Rice Grains with Potential Health Consequences for the Poorest Rice-Dependent Countries," *Science Advances* 4, no. 5 (May 2018), https://doi.org/10.1126/sciadv.aaq1012.

Drowning

59   **four feet of sea-level rise:** Brady Dennis and Chris Mooney, "Scientists Nearly Double Sea Level Rise Projections for 2100, Because of Antarctica," *The Washington Post*, March 30, 2016.

59   **by the end of the century:** Benjamin Strauss and Scott Kulp, "Extreme Sea Level Rise and the Stakes for America," Climate Central, April 26, 2017, www .climatecentral.org/news/extreme-sea-level-rise-stakes-for-america-21387.

59   **A radical reduction:** See the graphic "Surging Seas: 2°C Warming and Sea Level Rise" on the Climate Central website.

60   **Jeff Goodell runs through:** Jeff Goodell, *The Water Will Come: Rising Seas, Sinking Cities, and the Remaking of the Civilized World* (New York: Little, Brown, 2017), p. 13.

60   **Atlantis:** The historical basis, if any, for this legend remains a subject of debate and dispute, but for an overview (and the suggestion that the society was submerged by a volcano eruption on today's Santorini), see Willie Drye, "Atlantis," *National Geographic*, 2018.

60   **as much as 5 percent:** Jochen Hinkel et al., "Coastal Flood Damage and Adaptation Costs Under 21st Century Sea-Level Rise," *Proceedings of the National Academy of Sciences* (February 2014), https://doi.org/10.1073/ pnas.1222469111.

60   **Jakarta is one:** Mayuri Mei Lin and Rafki Hidayat, "Jakarta, the Fastest-Sinking City in the World," *BBC News*, August 13, 2018, www.bbc.com/ news/world-asia-44636934.

60   **China is evacuating:** Andrew Galbraith, "China Evacuates 127,000 People as Heavy Rains Lash Guangdong—Xinhua," Reuters, September 1, 2018, www.reuters.com/article/us-china-floods/china-evacuates-127000-people -as-heavy-rains-lash-guangdong-xinhua-idUSKCN1LH3BV.

61   **Much of the infrastructure:** Ramakrishnan Durairajan et al., "Lights Out: Climate Change Risk to Internet Infrastructure," *Proceedings of the Applied Networking Research Workshop* (July 16, 2018): pp. 9–15, https://doi .org/10.1145/3232755.3232775.

61   **nearly 311,000 homes:** Union of Concerned Scientists, "Underwater: Rising Seas, Chronic Floods, and the Implications for US Coastal Real Estate" (Cambridge, MA, 2018), p. 5, www.ucsusa.org/global-warming/global -warming-impacts/sea-level-rise-chronic-floods-and-us-coastal-real-estate -implications.

61   **$100 trillion *per year* by 2100:** University of Southampton, "Climate Change Threatens to Cause Trillions in Damage to World's Coastal Regions If They Do Not Adapt to Sea-Level Rise," February 4, 2014, www.southampton .ac.uk/news/2014/02/04-climate-change-threatens-damage-to-coastal -regions.page#.UvonXXewI2l.

61   **$14 trillion a year:** Svetlana Jevrejeva et al., "Flood Damage Costs Under the Sea Level Rise with Warming of 1.5 °C and 2 °C," *Environmental Research Letters* 13, no. 7 (July 2018), https://doi.org/10.1088/1748-9326/aacc76.

61   **continue for millennia:** Andrea Dutton et al., "Sea-Level Rise Due to Polar Ice-Sheet Mass Loss During Past Warm Periods," *Science* 349, no. 6244 (July 2015), https://doi.org/ 10.1126/science.aaa4019.

61   **two-degree scenario:** "Surging Seas," Climate Central.

61   **about 444,000 square miles:** Benjamin Strauss, "Coastal Nations, Megacities Face 20 Feet of Sea Rise," Climate Central, July 9, 2015, www.climatecentral.org/news/nations-megacities-face-20-feet-of-sea -level-rise-19217.

61   **the twenty cities most affected:** Ibid.

62   **flooding has quadrupled since 1980:** European Academies' Science Advisory Council, "New Data Confirm Increased Frequency of Extreme Weather Events, European National Science Academies Urge Further Action on Climate Change Adaptation," March 21, 2018, https://easac.eu/press-releases/ details/new-data-confirm-increased-frequency-of-extreme-weather-events -european-national-science-academies.

62   **by 2100 high-tide flooding:** National Oceanic and Atmospheric Administration, "Patterns and Projections of High Tide Flooding Along the US Coastline Using a Common Impact Threshold" (Silver Spring, MD, February 2018), p. ix, https://tidesandcurrents.noaa.gov/publications/techrpt86 _PaP_of_HTFlooding.pdf.

62   **affected 2.3 billion and killed 157,000:** United Nations Office for Disaster Risk Reduction, "The Human Cost of Weather Related Disasters 1995– 2015" (Geneva, 2015), p. 13, www.unisdr.org/2015/docs/climatechange/ COP21_WeatherDisastersReport_2015_FINAL.pdf.

62   **increase global rainfall to such a degree:** Sven N. Willner et al., "Adaptation Required to Preserve Future High-End River Flood Risk at Present Levels," *Science Advances* 4, no. 1 (January 2018), https://doi.org/10.1126/sciadv .aao1914.

62   **at risk of catastrophic inundation:** Oliver E. J. Wing et al., "Estimates of Present and Future Flood Risk in the Conterminous United States," *Environmental Research Letters* 13, no. 3 (February 2018), https://doi .org/10.1088/1748-9326/aaac65.

63   **floods in South Asia killed 1,200:** Oxfam International, "43 Million Hit by South Asia Floods: Oxfam Is Responding," August 31, 2017, www.oxfam .org/en/pressroom/pressreleases/2017-08-31/43-million-hit-south-asia -floods-oxfam-responding.

63   **António Guterres, the secretary-general:** United Nations Secretary-General, "Secretary-General's Press Encounter on Climate Change [with Q&A]," March 29, 2018, www.un.org/sg/en/content/sg/press-encounter/2018-03-29/ secretary-generals-press-encounter-climate-change-qa.

63   **eight times the entire global population:** U.S. Census Bureau, "Historical Estimates of World Population," www.census.gov/data/tables/time-series/ demo/international-programs/historical-est-worldpop.html.

63   **Noah's Ark story:** There are a number of theories about historical flood events that may have inspired the biblical story, but this popular one was presented

at length in William Ryan and Walter Pitman, *Noah's Flood: The New Scientific Discoveries About the Event That Changed History* (New York: Simon & Schuster, 2000).

63   **700,000 Rohingya refugees:** Michael Schwirtz, "Besieged Rohingya Face 'Crisis Within the Crisis': Deadly Floods," *The New York Times*, February 13, 2018.

63   **When the Paris Agreement was drafted:** Meehan Crist, "Besides, I'll Be Dead," *London Review of Books*, February 22, 2018, www.lrb.co.uk/v40/n04/meehan-crist/besides-ill-be-dead.

64   **"sunny day flooding":** Jim Morrison, "Flooding Hot Spots: Why Seas Are Rising Faster on the US East Coast," *Yale Environment 360*, April 24, 2018, https://e360.yale.edu/features/flooding-hot-spots-why-seas-are-rising-faster-on-the-u.s.-east-coast.

64   **things accelerating faster:** Andrew Shepherd, Helen Amanda Fricker, and Sinead Louise Farrell, "Trends and Connections Across the Antarctic Cryosphere," *Nature* 558 (2018): pp. 223–32.

63   **melt rate of the Antarctic:** University of Leeds, "Antarctica Ramps Up Sea Level Rise," June 13, 2018, www.leeds.ac.uk/news/article/4250/antarctica_ramps_up_sea_level_rise.

64   **49 billion tons of ice each year:** Chris Mooney, "Antarctic Ice Loss Has Tripled in a Decade. If That Continues, We Are in Serious Trouble," *The Washington Post*, June 13, 2018.

64   **several meters over fifty years:** James Hansen et al., "Ice Melt, Sea Level Rise, and Superstorms: Evidence from Paleoclimate Data, Climate Modeling, and Modern Observations That 2°C Global Warming Could Be Dangerous," *Atmospheric Chemistry and Physics* 16 (2016): pp. 3761–812, https://doi.org/10.5194/acp-16-3761-2016.

64   **13,000 square miles:** University of Maryland, "Decades of Satellite Monitoring Reveal Antarctic Ice Loss," June 13, 2018, https://cmns.umd.edu/news-events/features/4156.

64   **determined by what human action:** Hayley Dunning, "How to Save Antarctica (and the Rest of Earth Too)," Imperial College London, June 13, 2018, www.imperial.ac.uk/news/186668/how-save-antarctica-rest-earth.

64   **never before observed in human history:** Richard Zeebe et al., "Anthropogenic Carbon Release Rate Unprecedented During the Past 66 Million Years," *Nature Geoscience* 9 (March 2016): pp. 325–29, https://doi.org//10.1038/ngeo2681.

64   **"damage mechanics":** C. P. Borstad et al., "A Damage Mechanics Assessment of the Larsen B Ice Shelf Prior to Collapse: Toward a Physically-Based Calving Law," *Geophysical Research Letters* 39 (September 2012), https://doi.org/10.1029/2012GL053317.

65   **around ten times faster:** Sarah Griffiths, "Global Warming Is Happening 'Ten Times Faster than at Any Time in the Earth's History,' Climate Experts Claim," *The Daily Mail*, August 2, 2013. See also Melissa Davey, "Humans Causing Climate to Change 170 Times Faster than Natural Forces," *The Guardian*, February 12, 2017; this estimate for a rate of warming 170

times faster came from Owen Gaffney and Will Steffen, "The Anthropocene Equation," *The Anthropocene Review*, February 10, 2017, https://doi.org/10.1177/2053019616688022.

65   **the average American emits:** Dirk Notz and Julienne Stroeve, "Observed Arctic Sea-Ice Loss Directly Follows Anthropogenic CO_2 Emission," *Science*, November 3, 2016. See also Robinson Meyer, "The Average American Melts 645 Square Feet of Arctic Ice Every Year," *The Atlantic*, November 3, 2016. And see also Ken Caldeira, "How Much Ice Is Melted by Each Carbon Dioxide Emission?" March 24, 2018, https://kencaldeira.wordpress.com/2018/03/24/how-much-ice-is-melted-by-each-carbon-dioxide-emission.

65   **1.2 degrees of global warming:** Sebastian H. Mernild, "Is 'Tipping Point' for the Greenland Ice Sheet Approaching?" *Aktuel Naturvidenskab*, 2009, http://mernild.com/onewebmedia/2009.AN%20Mernild4.pdf.

65   **raise sea levels six meters:** National Snow and Ice Data Center, "Quick Facts on Ice Sheets," https://nsidc.org/cryosphere/quickfacts/icesheets.html.

65   **West Antarctic and Greenland ice sheets:** Patrick Lynch, "The 'Unstable' West Antarctic Ice Sheet: A Primer," NASA, May 12, 2014, www.nasa.gov/jpl/news/antarctic-ice-sheet-20140512.

65   **a billion tons of ice:** UMassAmherst College of Engineering, "Gleason Participates in Groundbreaking Greenland Research That Makes Front Page of *New York Times*," January 2017, https://engineering.umass.edu/news/gleason-participates-groundbreaking-greenland-research-that-makes-front-page-new-york-times.

65   **raise global sea levels ten to twenty feet:** Jonathan L. Bamber, "Reassessment of the Potential Sea-Level Rise from a Collapse of the West Antarctic Ice Sheet," *Science* 324, no. 5929 (May 2009): pp. 901–3, https://doi.org/10.1126/science.1169335.

65   **eighteen billion tons of ice:** Alejandra Borunda, "We Know West Antarctica Is Melting. Is the East in Danger, Too?" *National Geographic*, August 10, 2018.

66   **permafrost contains up to 1.8 trillion:** NASA Science, "Is Arctic Permafrost the 'Sleeping Giant' of Climate Change?" June 24, 2013, https://science.nasa.gov/science-news/science-at-nasa/2013/24jun_permafrost.

66   **one *Nature* paper found that:** Katey Walter Anthony et al., "21st-Century Modeled Permafrost Carbon Emissions Accelerated by Abrupt Thaw Beneath Lakes," *Nature Communications* 9, no. 3262 (August 2018), https://doi.org/10.1038/s41467-018-05738-9. See also Ellen Gray, "Unexpected Future Boost of Methane Possible from Arctic Permafrost," NASA Climate, August 20, 2018, https://climate.nasa.gov/news/2785/unexpected-future-boost-of-methane-possible-from-arctic-permafrost.

66   **"abrupt thawing":** Anthony, "21st-Century Modeled Permafrost Carbon Emissions," https://doi.org/10.1038/s41467-018-05738-9.

66   **Atmospheric methane levels have risen:** "What Is Behind Rising Levels of Methane in the Atmosphere?" NASA Earth Observatory, January 11, 2018, https://earthobservatory.nasa.gov/images/91564/what-is-behind-rising-levels-of-methane-in-the-atmosphere.

66   **Arctic lakes could possibly double:** Anthony, "21st-Century Modeled Permafrost Carbon Emissions," https://doi.org/10.1038/s41467-018-05738-9.

67   **between 37 and 81 percent by 2100:** IPCC, *Climate Change 2013: The Physical Science Basis—Summary for Policymakers* (Geneva, October 2013), p. 23, www.ipcc.ch/pdf/assessment-report/ar5/wg1/WGIAR5_SPM_brochure_en.pdf.

67   **as quickly as the 2020s:** Kevin Schaeffer et al., "Amount and Timing of Permafrost Release in Response to Climate Warming," *Tellus B*, January 24, 2011.

67   **a hundred billion tons:** Ibid.

67   **a massive warming equivalent:** Peter Wadhams, "The Global Impacts of Rapidly Disappearing Arctic Sea Ice," *Yale Environment 360*, September 26, 2016, https://e360.yale.edu/features/as_arctic_ocean_ice_disappears_global_climate_impacts_intensify_wadhams.

68   **at least fifty meters:** David Archer, *The Long Thaw: How Humans Are Changing the Next 100,000 Years of Earth's Climate* (Princeton, NJ: Princeton University Press, 2016).

68   **The U.S. Geological Survey:** Jason Treat et al., "What the World Would Look Like If All the Ice Melted," *National Geographic*, September 2013.

68   **more than 97 percent of Florida:** Benjamin Strauss, Scott Kulp, and Peter Clark, "Can You Guess What America Will Look Like in 10,000 Years? A Quiz," *The New York Times*, April 20, 2018, www.nytimes.com/interactive/2018/04/20/sunday-review/climate-flood-quiz.html.

68   **Manaus, the capital:** Treat, "What the World Would Look Like."

69   **More than 600 million:** Gordon McGranahan et al., "The Rising Tide: Assessing the Risks of Climate Change and Human Settlements in Low Elevation Coastal Zones," *Environment and Urbanization* 19, no. 1 (April 2007): pp. 17–27, https://doi.org/10.1177//0956247807076960.

Wildfire

70   **Thomas Fire, the worst:** CalFire, "Incident Information: Thomas Fire," March 28, 2018, http://cdfdata.fire.ca.gov/incidents/incidents_details_info?incident_id=1922.

70   **"15% contained":** CalFire, "Thomas Fire Incident Update," December 11, 2017, http://cdfdata.fire.ca.gov/pub/cdf/images/incidentfile1922_3183.pdf.

70   **"Los Angeles Notebook":** Joan Didion, *Slouching Towards Bethlehem* (New York: Farrar, Straus & Giroux, 1968).

71   **Five of the twenty worst fires:** CalFire, "Top 20 Most Destructive California Wildfires," August 20, 2018, www.fire.ca.gov/communications/downloads/fact_sheets/Top20_Destruction.pdf.

71   **1,240,000 acres:** CalFire, "Incident Information: 2017," January 24, 2018, http://cdfdata.fire.ca.gov/incidents/incidents_stats?year=2017.

71   **172 fires broke out:** California Board of Forestry and Fire Protection, "October 2017 Fire Siege," January 2018, http://bofdata.fire.ca.gov/board_business/

binder_materials/2018/january_2018_meeting/full/full_14_presentation
_october_2017_fire_siege.pdf.

71   **One couple survived:** Robin Abcarian, "They Survived Six Hours in a Pool as a Wildfire Burned Their Neighborhood to the Ground," *Los Angeles Times*, October 12, 2017.

71   **only the husband who emerged:** Erin Allday, "Wine Country Wildfires: Huddled in Pool amid Blaze, Wife Dies in Husband's Arms," *SF Gate*, January 25, 2018.

71   **more than two thousand square miles:** CalFire, "Incident Information: 2018," January 24, 2018, http://cdfdata.fire.ca.gov/incidents/incidents_stats ?year=2018.

72   **smoke blanketed almost half the country:** Megan Molteni, "Wildfire Smoke Is Smothering the US—Even Where You Don't Expect It," *Wired*, August 14, 2018.

72   **in British Columbia:** Estefania Duran, "B.C. Year in Review 2017: Wildfires Devastate the Province like Never Before," *Global News*, December 25, 2017, https://globalnews.ca/news/3921710/b-c-year-in-review-2017-wildfires.

72   **L.A. has always been:** Mike Davis, *City of Quartz: Excavating the Future in Los Angeles* (London: Verso, 1990).

73   **burned the state's wine crop:** Tiffany Hsu, "In California Wine Country, Wildfires Take a Toll on Vintages and Tourism," *The New York Times*, October 10, 2017.

73   **Getty Museum:** Jessica Gelt, "Getty Museum Closes Because of Fire, but 'The Safest Place for the Art Is Right Here,' Spokesman Says," *Los Angeles Times*, December 6, 2017.

74   **wildfire season in the western United States:** "Climate Change Indicators: U.S. Wildfires," WX Shift, http://wxshift.com/climate-change/climate-indicators/us-wildfires.

74   **nearly 20 percent:** W. Matt Jolly et al., "Climate-Induced Variations in Global Wildfire Danger from 1979 to 2013," *Nature Communications* 6, no. 7537 (July 2015), https://doi.org/10.1038/ncomms8537.

74   **By 2050, destruction:** Joseph Romm, *Climate Change: What Everyone Needs to Know* (New York: Oxford University Press, 2016), p. 47.

74   **ten million acres were burned:** National Interagency Fire Center, "Total Wildland Fires and Acres (1926-2017)," www.nifc.gov/fireInfo/fireInfo_stats_totalFires.html.

74   **"We don't even call it":** Melissa Pamer and Elizabeth Espinosa, "'We Don't Even Call It Fire Season Anymore . . . It's Year Round': Cal Fire," KTLA 5, December 11, 2017, https://ktla.com/2017/12/11/we-dont-even-call-it-fire-season-anymore-its-year-round-cal-fire.

74   **soot and ash they give off:** William Finnegan, "California Burning," *New York Review of Books*, August 16, 2018.

75   **dozens of guests tried to escape:** Jason Horowitz, "As Greek Wildfire Closed In, a Desperate Dash Ended in Death," *The New York Times*, July 24, 2018.

75   **Great Flood of 1862:** Daniel L. Swain et al., "Increasing Precipitation Vola-
       tility in Twenty-First-Century California," *Nature Climate Change* 8 (April
       2018): pp. 427–33, https://doi.org/10.1038/s41558-018-0140-y.

75   **globally, between 260,000 and 600,000:** Fay H. Johnston et al., "Estimated
       Global Mortality Attributable to Smoke from Landscape Fires," *Environ-
       mental Health Perspectives* 120, no. 5 (May 2012), https://doi.org/10.1289/
       ehp.1104422.

75   **Canadian fires have:** George E. Le et al., "Canadian Forest Fires and the
       Effects of Long-Range Transboundary Air Pollution on Hospitalizations
       Among the Elderly," *ISPRS International Journal of Geo-Information* 3 (May
       2014): pp. 713–31, https://doi.org/ 10.3390/ijgi3020713.

75   **42 percent spike in hospital visits:** C. Howard et al., "SOS: Summer of
       Smoke—A Mixed-Methods, Community-Based Study Investigating the
       Health Effects of a Prolonged, Severe Wildfire Season on a Subarctic Pop-
       ulation," *Canadian Journal of Emergency Medicine* 19 (May 2017): p. S99,
       https://doi.org/10.1017/cem.2017.264.

75   **"One of the strongest emotions":** Sharon J. Riley, "'The Lost Summer': The
       Emotional and Spiritual Toll of the Smoke Apocalypse," *The Narwhal*, Au-
       gust 21, 2018, https://thenarwhal.ca/the-lost-summer-the-emotional-and
       -spiritual-toll-of-the-smoke-apocalypse.

76   **Peatland fires in Indonesia:** Susan E. Page et al., "The Amount of Carbon
       Released from Peat and Forest Fires in Indonesia During 1997," *Nature* 420
       (November 2002): pp. 61–65, https://doi.org/10.1038/nature01131. For a
       picture of how peatland emissions will change going forward, see Angela
       V. Gallego-Sala et al., "Latitudinal Limits to the Predicted Increase of the
       Peatland Carbon Sink with Warming," *Nature Climate Change* 8 (2018):
       pp. 907–13.

76   **In California, a single wildfire:** David R. Baker, "Huge Wildfires Can Wipe
       Out California's Greenhouse Gas Gains," *San Francisco Chronicle*, Novem-
       ber 22, 2017.

76   **its second "hundred-year drought":** Joe Romm, "Science: Second '100-
       Year' Amazon Drought in Five Years Caused Huge CO_2 Emissions. If
       This Pattern Continues, the Forest Would Become a Warming Source,"
       ThinkProgress, February 8, 2011, https://thinkprogress.org/science-second
       -100-year-amazon-drought-in-5-years-caused-huge-co2-emissions-if-this
       -pattern-7036a9074098.

76   **the trees of the Amazon:** Roel J. W. Brienen et al., "Long-Term Decline of
       the Amazon Carbon Sink," *Nature*, March 2015.

76   **A group of Brazilian scientists:** Aline C. Soterroni et al., "Fate of the Ama-
       zon Is on the Ballot in Brazil's Presidential Election," *Monga Bay*, Octo-
       ber 17, 2018, https://news.mongabay.com/2018/10/fate-of-the-amazon-is-on
       -the-ballot-in-brazils-presidential-election-commentary/.

77   **deforestation accounts for about 12 percent:** G. R. van der Werf et al.,
       "CO_2 Emissions from Forest Loss," *Nature Geoscience* 2 (November 2009):
       pp. 737–38, https://doi.org/10.1038/ngeo671.

77   **as much as 25 percent:** Bob Berwyn, "How Wildfires Can Affect Climate
       Change (and Vice Versa)," *Inside Climate News*, August 23, 2018, https://

insideclimatenews.org/news/23082018/extreme-wildfires-climate-change
-global-warming-air-pollution-fire-management-black-carbon-co2.

77   **ability of forest soils to absorb:** Daisy Dunne, "Methane Uptake from Forest
Soils Has 'Fallen by 77% in Three Decades,'" *Carbon Brief*, August 6, 2018,
www.carbonbrief.org/methane-uptake-from-forest-soils-has-fallen-77-per
-cent-three-decades.

77   **an additional 1.5 degrees Celsius:** Natalie M. Mahowald et al., "Are the
Impacts of Land Use on Warming Underestimated in Climate Policy?"
Environmental Research Letters 12, no. 9 (September 2017), https://doi
.org/10.1088/1748-9326/aa836d.

77   **30 percent of emissions:** Quentin Lejeune et al., "Historical Deforestation
Locally Increased the Intensity of Hot Days in Northern Mid-Latitudes,"
Nature Climate Change 8 (April 2018), pp. 386–90, https://doi.org/10.1038/
s41558-018-0131-z.

77   **twenty-seven additional cases:** Leonardo Suveges Moreira Chaves et al.,
"Abundance of Impacted Forest Patches Less than 5 km² Is a Key Driver of
the Incidence of Malaria in Amazonian Brazil," *Scientific Reports* 8, no. 7077
(May 2018), https://doi.org/10.1038/s41598-018-25344-5.

Disasters No Longer Natural

78   **tornadoes will strike much more frequently:** Francesco Fiondella, "Extreme
Tornado Outbreaks Have Become More Common," International Research
Institute for Climate and Society, Columbia University, March 2, 2016,
https://iri.columbia.edu/news/tornado-outbreaks.

78   **their trails of destruction could grow:** Joseph Romm, *Climate Change: What
Everyone Needs to Know* (New York: Oxford University Press, 2016), p. 69.

79   **three major hurricanes:** Congressional Research Service, *The National Hurricane Center and Forecasting Hurricanes: 2017 Overview and 2018 Outlook*
(Washington, D.C., August 23, 2018), https://fas.org/sgp/crs/misc/R45264
.pdf.

79   **dropping on Houston:** Javier Zarracina and Brian Resnick, "All the Rain
That Hurricane Harvey Dumped on Texas and Louisiana, in One Massive
Water Drop," *Vox*, September 1, 2017.

79   **record-breaking summer of 2018:** Jason Samenow, "Red Hot Planet: This
Summer's Punishing and Historic Heat in Seven Charts and Maps," *The
Washington Post*, August 17, 2018.

79   **In 1850, the area had 150 glaciers:** U.S. Geological Survey, "Retreat of Glaciers in Glacier National Park," April 6, 2016, www.usgs.gov/centers/norock/
science/retreat-glaciers-glacier-national-park.

80   **Already, storms have doubled since 1980:** European Academies' Science
Advisory Council, "New Data Confirm Increased Frequency of Extreme
Weather Events, European National Science Academies Urge Further Action on Climate Change Adaptation," March 21, 2018, https://easac.eu/
press-releases/details/new-data-confirm-increased-frequency-of-extreme
-weather-events-european-national-science-academies.

80   **New York City will suffer:** Andra J. Garner et al., "Impact of Climate Change on New York City's Coastal Flood Hazard: Increasing Flood Heights from the Preindustrial to 2300 CE," *Proceedings of the National Academy of Sciences* (September 2017), https://doi.org/10.1073/pnas.1703568114.

80   **more intense rainstorms:** U.S. Global Change Research Program, *2014 National Climate Assessment* (Washington, D.C., 2014), https://nca2014.globalchange.gov/report/our-changing-climate/heavy-downpours-increasing.

80   **In the Northeast:** U.S. Global Change Research Program, "Observed Change in Very Heavy Precipitation," September 19, 2013, https://data.globalchange.gov/report/nca3/chapter/our-changing-climate/figure/observed-change-in-very-heavy-precipitation-2.

80   **The island of Kauai:** National Weather Service, "April 2018 Precipitation Summary," May 4, 2018, www.prh.noaa.gov/hnl/hydro/pages/apr18sum.php.

80   **the damages from quotidian thunderstorms:** Alyson Kenward and Urooj Raja, "Blackout: Extreme Weather, Climate Change and Power Outages," Climate Central (Princeton, NJ, 2014), p. 4, http://assets.climatecentral.org/pdfs/PowerOutages.pdf.

80   **When Hurricane Irma first emerged:** Joe Romm, "The Case for a Category 6 Rating for Super-Hurricanes like Irma," *ThinkProgress*, September 6, 2017, https://thinkprogress.org/category-six-hurricane-irma-62cfdfdd93cb.

81   **flooding its agricultural lands:** Frances Robles and Luis Ferré-Sadurní, "Puerto Rico's Agriculture and Farmers Decimated by Maria," *The New York Times*, September 24, 2017.

81   **"We're getting some intimations":** This was a comment Wark made on Twitter: https://twitter.com/mckenziewark/status/913382357230645248.

81   **seventeen times more often:** Ning Lin et al., "Hurricane Sandy's Flood Frequency Increasing from Year 1800 to 2100," *Proceedings of the National Academy of the Sciences,* October 2016.

81   **Katrina-level hurricanes are expected:** Aslak Grinsted et al., "Projected Atlantic Hurricane Surge Threat from Rising Temperatures," *Proceedings of the National Academy of Sciences* (March 2013), https://doi.org/10.1073/pnas.1209980110.

81   **Looking globally, researchers have found:** Greg Holland and Cindy L. Bruyère, "Recent Intense Hurricane Response to Global Climate Change," *Climate Dynamics* 42, no. 3–4 (February 2014): pp. 617–27, https://doi.org/10.1007/s00382-013-1713-0.

81   **Between just 2006 and 2013, the Philippines:** Food and Agriculture Organization, "The Impact of Disasters on Agriculture and Food Security" (Rome, 2015), p. xix, https://reliefweb.int/sites/reliefweb.int/files/resources/a-i5128e.pdf.

82   **typhoons have intensified:** Wei Mei and Shang-Ping Xie, "Intensification of Landfalling Typhoons over the Northwest Pacific Since the Late 1970s," *Nature Geoscience* 9 (September 2016): pp. 753–57, https://doi.org/10.1038/NGEO2792.

82   **By 2070, Asian megacities:** Linda Poon, "Climate Change Is Testing Asia's Megacities," CityLab, October 9, 2018, www.citylab.com/
environment/2018/10/asian-megacities-vs-tomorrows-typhoons/572062.

82   **the more intense the blizzards:** Judah Cohen et al., "Warm Arctic Epi-
sodes Linked with Increased Frequency of Extreme Winter Weather in the
United States," *Nature Communications* 9, no. 869 (March 2018): https://doi
.org/10.1038/s41467-018-02992-9.

82   **758 tornadoes:** NOAA National Centers for Environmental Information,
"State of the Climate: Tornadoes for April 2011," May 2011, www.ncdc
.noaa.gov/sotc/tornadoes/201104.

83   **40 percent by 2010:** Noah S. Diffenbaugh et al., "Robust Increases in Severe
Thunderstorm Environments in Response to Greenhouse Forcing," *Proceedings of the National Academy of Sciences* 110, no. 41 (October 2013): pp. 16361–
66, https://doi.org/10.1073/pnas.1307758110.

83   **$725 billion:** Keith Porter et al., "Overview of the ARkStorm Scenario," U.S.
Geological Survey, January 2011, https://pubs.usgs.gov/of/2010/1312.

83   **a cloud of "unbearable" smells:** Emily Atkin, "Minutes: 'Unbearable'
Petrochemical Smells Are Reportedly Drifting into Houston," *The New Republic*, August 2017.

83   **nearly half a billion gallons:** Frank Bajak and Lise Olsen, "Silent Spills,"
Houston Chronicle, May 2018.

84   **the city had already been knocked:** Kevin Litten, "16 New Orleans Pumps,
Not 14, Were Down Saturday and Remain Out: Officials," *The Times-
Picayune*, August 10, 2017.

84   **the 2000 population of 480,000:** Elizabeth Fussell, "Constructing New Orleans, Constructing Race: A Population History of New Orleans," *The Journal of American History* 94, no. 3 (December 2007), pp. 846–55, www.jstor
.org/stable/25095147.

84   **as low as 230,000:** Allison Plyer, "Facts for Features: Katrina Impact," The
Data Center, August 26, 2016, www.datacenterresearch.org/data-resources/
katrina/facts-for-impact.

84   **One of the fastest-growing cities:** U.S. Census Bureau, "The South Is
Home to 10 of the 15 Fastest-Growing Large Cities," May 25, 2017, www
.census.gov/newsroom/press-releases/2017/cb17-81-population-estimates
-subcounty.html.

84   **included the fastest-growing suburb:** Amy Newcomb, "Census Bureau Reveals Fastest-Growing Large Cities," U.S. Census Bureau, 2018.

84   **more than five times as many residents:** U.S. Census Bureau figures.

84   **brought there by the oil business:** John Schwartz, "Exxon Misled the Public
on Climate Change, Study Says," *The New York Times*, August 23, 2017.

84   **Lower Ninth Ward:** Greg Allen, "Ghosts of Katrina Still Haunt New
Orleans' Shattered Lower Ninth Ward," NPR, August 3, 2015, www.npr
.org/2015/08/03/427844717/ghosts-of-katrina-still-haunt-new-orleans
-shattered-lower-ninth-ward.

85   **the entire coastline of Louisiana:** Kevin Sack and John Schwartz, "Left to Louisiana's Tides, a Village Fights for Time," *The New York Times*, February 24, 2018, www.nytimes.com/interactive/2018/02/24/us/jean-lafitte-floodwaters .html.

85   **2,000 square miles already gone:** Bob Marshall, Brian Jacobs, and Al Shaw, "Losing Ground," *ProPublica*, August 28, 2014, http://projects.propublica .org/louisiana.

85   **2018 road budget:** Jeff Goodell, "Welcome to the Age of Climate Migration," *Rolling Stone*, February 4, 2018.

85   **islanders arrived in Florida:** John D. Sutter and Sergio Hernandez, "'Exodus' from Puerto Rico: A Visual Guide," CNN, February 21, 2018, www .cnn.com/2018/02/21/us/puerto-rico-migration-data-invs/index.html.

Freshwater Drain

86   **Seventy-one percent of the planet:** USGS Water Science School, "How Much Water Is There on, in, and Above the Earth?" U.S. Geological Survey, December 2, 2016, https://water.usgs.gov/edu/earthhowmuch.html.

86   **Barely more than 2 percent:** USGS Water Science School, "The World's Water," U.S. Geological Survey, December 2, 2016, https://water.usgs.gov/ edu/earthwherewater.html.

86   **only 0.007 percent of the planet's water:** "Freshwater Crisis," *National Geographic*.

86   **Globally, between 70 and 80 percent:** Tariq Khokhar, "Chart: Globally, 70% of Freshwater Is Used for Agriculture," World Bank Data Blog, March 22, 2017, https://blogs.worldbank.org/opendata/chart-globally-70 -freshwater-used-agriculture.

87   **twenty liters of water each day:** "Water Consumption in Africa," Institute Water for Africa, https://water-for-africa.org/en/water-consumption/ar-ticles/water-consumption-in-africa.html.

87   **less than half of what water organizations:** UN-Water Decade Programme on Advocacy and Communication and Water Supply and Sanitation Collaborative Council, "The Human Right to Water and Sanitation," www.un.org/ waterforlifedecade/pdf/human_right_to_water_and_sanitation_media _brief.pdf.

87   **global water demand is expected:** "Half the World to Face Severe Water Stress by 2030 Unless Water Use Is 'Decoupled' from Economic Growth, Says International Resource Panel," United Nations Environment Programme, March 21, 2016, www.unenvironment.org/news-and-stories/ press-release/half-world-face-severe-water-stress-2030-unless-water-use -decoupled.

87   **loss of 16 percent of freshwater:** "Water Audits and Water Loss Control for Public Water Systems," Environmental Protection Agency, July 2013, www .epa.gov/sites/production/files/2015-04/documents/epa816f13002.pdf.

87   **in Brazil, the estimate is 40 percent:** "Treated Water Loss Is Still High in Brazil," World Water Forum, November 21, 2017, http://8.worldwaterforum .org/en/news/treated-water-loss-still-high-brazil.

87   **a tool of inequality:** In 2018, it was revealed that Harvard had aggressively bought up California vineyards for the water underground.

87   **2.1 billion people around the world:** "2.1 Billion People Lack Safe Drinking Water at Home, More than Twice as Many Lack Safe Sanitation," World Health Organization, July 12, 2017, www.who.int/news-room/detail/12-07 -2017-2-1-billion-people-lack-safe-drinking-water-at-home-more-than -twice-as-many-lack-safe-sanitation.

87   **4.5 billion don't have safely managed water:** Ibid.

87   **Half of the world's population:** M. Huss et al., "Toward Mountains Without Permanent Snow and Ice," *Earth's Future* 5, no. 5 (May 2017): pp. 418–35, https://doi.org/10.1002/2016EF000514.

87   **the glaciers of the Himalayas:** P. D. A. Kraaijenbrink, "Impact of a Global Temperature Rise of 1.5 Degrees Celsius on Asia's Glaciers," *Nature* 549 (September 2017): pp. 257–60, https://doi.org/ 10.1038/nature23878.

87   **At four degrees, the snow-capped Alps:** Mark Lynas, *Six Degrees: Our Future on a Hotter Planet* (Washington, D.C.: National Geographic Society, 2008), p. 202.

87   **70 percent less snow:** Christoph Marty et al., "How Much Can We Save? Impact of Different Emission Scenarios on Future Snow Cover in the Alps," *The Cryosphere*, 2017.

88   **250 million Africans:** United Nations Framework Convention on Climate Change, "Climate Change: Impacts, Vulnerabilities and Adaptation in De- veloping Countries" (New York, 2007), p. 5, https://unfccc.int/resource/ docs/publications/impacts.pdf.

88   **a billion people in Asia:** Charles Fant et al., "Projections of Water Stress Based on an Ensemble of Socioeconomic Growth and Climate Change Sce- narios: A Case Study in Asia," *PLOS One* 11, no. 3 (March 2016), https://doi .org/10.1371/journal.pone.0150633.

88   **freshwater availability in cities:** World Bank, "High and Dry: Climate Change, Water, and the Economy" (Washington, D.C., 2016), p. vi.

88   **five billion people:** UN Water, "The United Nations World Water Develop- ment Report 2018: Nature-Based Solutions for Water" (Paris, 2018), p. 3, http://unesdoc.unesco.org/images/0026/002614/261424e.pdf.

88   **boomtown Phoenix:** Marcello Rossi, "Desert City Phoenix Mulls Ways to Quench Thirst of Sprawling Suburbs," *Thomson Reuters Foundation News*, June 7, 2018, news.trust.org/item/20180607120002-7kwzq.

88   **even London is beginning to worry:** Edoardo Borgomeo, "Will London Run Out of Water?" *The Conversation*, May 24, 2018, https://theconversation .com/will-london-run-out-of-water-97107.

88   **"high to extreme water stress":** NITI Aayog, *Composite Water Manage- ment Index: A Tool for Water Management* (June 2018), p. 15, www.niti.gov .in/writereaddata/files/document_publication/2018-05-18-Water-index -Report_vS6B.pdf.

88   **water availability in Pakistan:** Rina Saeed Khan, "Water Pressures Rise in Pakistan as Drought Meets a Growing Population," Reuters, June 14, 2018, https://af.reuters.com/article/commoditiesNews/idAFL5N1T7502.

88   **the Aral Sea:** NASA Earth Observatory, "World of Change: Shrinking Aral Sea," https://earthobservatory.nasa.gov/WorldOfChange/AralSea.

88   **Lake Poopó:** NASA Earth Observatory, "Bolivia's Lake Poopó Disappears," January 23, 2016, https://earthobservatory.nasa.gov/images/87363/bolivias-lake-poopo-disappears.

88   **Lake Urmia:** Amir AghaKouchak et al., "Aral Sea Syndrome Desiccates Lake Urmia: Call for Action," *Journal of Great Lakes Research* 41, no. 1 (March 2015): pp. 307–11, https://doi.org/10.1016/j.jglr.2014.12.007.

88   **Lake Chad:** "Africa's Vanishing Lake Chad," *Africa Renewal* (April 2012), www.un.org/africarenewal/magazine/april-2012/africa%E2%80%99s-vanishing-lake-chad.

88   **warmwater-friendly bacteria:** Boqiang Qin et al., "A Drinking Water Crisis in Lake Taihu, China: Linkage to Climatic Variability and Lake Management," *Environmental Management* 45, no. 1 (January 2010): pp. 105–12, https://doi.org/10.1007/s00267-009-9393-6.

88   **Lake Tanganyika:** Jessica E. Tierney et al., "Late-Twentieth-Century Warming in Lake Tanganyika Unprecedented Since AD 500," *Nature Geoscience* 3 (May 2010): pp. 422–25, https://doi.org/10.1038/ngeo865. See also, for instance, Clea Broadhurst, "Global Warming Depletes Lake Tanganikya's Fish Stocks," RFI, August 9, 2016, http://en.rfi.fr/africa/20160809-global-warming-responsible-decline-fish-lake-tanganyika.

89   **16 percent of the world's natural methane:** E. J. S. Emilson et al., "Climate-Driven Shifts in Sediment Chemistry Enhance Methane Production in Northern Lakes," *Nature Communications* 9, no. 1801 (May 2018), https://doi.org/10.1038/s41467-018-04236-2. See also David Bastviken et al., "Methane Emissions from Lakes: Dependence of Lake Characteristics, Two Regional Assessments, and a Global Estimate," *Global Biogeochemical Cycles* 18 (2004), https://doi.org/10.1029/2004GB002238.

89   **could double those emissions:** "Greenhouse Gas 'Feedback Loop' Discovered in Freshwater Lakes," University of Cambridge, May 4, 2018, www.cam.ac.uk/research/news/greenhouse-gas-feedback-loop-discovered-in-freshwater-lakes.

89   **aquifers already supply:** USGS Water Science School, "Groundwater Use in the United States," U.S. Geological Survey, June 26, 2018, https://water.usgs.gov/edu/wugw.html.

89   **wells that used to draw water:** Brian Clark Howard, "California Drought Spurs Groundwater Drilling Boom in Central Valley," *National Geographic*, August 16, 2014.

89   **lost twelve cubic miles:** Kevin Wilcox, "Aquifers Depleted in Colorado River Basin," *Civil Engineering*, August 5, 2014, www.asce.org/magazine/20140805-aquifers-depleted-in-colorado-river-basin.

89   **Ogallala Aquifer:** Sandra Postel, "Drought Hastens Groundwater Depletion in the Texas Panhandle," *National Geographic*, July 24, 2014.

89   **expected to drain by 70 percent:** Kansas State University, "Study Forecasts Future Water Levels of Crucial Agricultural Aquifer," *K-State News*, August 26, 2013, www.k-state.edu/media/newsreleases/aug13/groundwater82613

.html. See also David R. Steward et al., "Tapping Unsustainable Groundwater Stores for Agricultural Production in the High Plains Aquifer of Kansas, Projections to 2110," *Proceedings of the National Academy of Sciences of the United States of America* 110. no. 37 (September 2013), pp. E3477–86, https://doi.org/10.1073/pnas.1220351110.

89   **twenty-one cities:** NITI Aayog, *Composite Water Management Index*, p. 22, www.niti.gov.in/writereaddata/files/document_publication/2018-05-18-Water-index-Report_vS6B.pdf.

89   **The first Day Zero:** City of Cape Town, "Day Zero: When Is It, What Is It, and How Can We Avoid It?" November 15, 2017.

89   **In a memorable first-person account:** Adam Welz, "Letter from a Bed in Cape Town," *Sierra*, February 12, 2018, www.sierraclub.org/sierra/letter-bed-cape-town-drought-day-zero.

90   **in arid Utah:** Mark Milligan, "Glad You Asked: Does Utah Really Use More Water than Any Other State?" Utah Geological Survey, https://geology.utah.gov/map-pub/survey-notes/glad-you-asked/does-utah-use-more-water.

90   **South Africa had nine million people:** UNESCO, *Water: A Shared Responsibility—The United Nations World Water Development Report 2* (Paris, 2006), p. 502, http://unesdoc.unesco.org/images/0014/001454/145405e.pdf#page=519.

91   **to produce the nation's wine crop:** Stephen Leahy, "From Not Enough to Too Much, the World's Water Crisis Explained," *National Geographic*, March 22, 2018.

91   **total urban consumption:** Public Policy Institute for California, "Water Use in California," July 2016, www.ppic.org/publication/water-use-in-california.

91   **limiting water use to twelve hours:** Jon Gerberg, "A Megacity Without Water: São Paulo's Drought," *Time*, October 13, 2015.

91   **aggressive rationing system:** Simon Romero, "Taps Start to Run Dry in Brazil's Largest City," *The New York Times*, February 16, 2015.

91   **barge in drinking water from France:** Graham Keeley, "Barcelona Forced to Import Emergency Water," *The Guardian*, May 14, 2008.

91   **"millennium drought":** "Recent Rainfall, Drought and Southern Australia's Long-Term Rainfall Decline," Australian Government Bureau of Meteorology, April 2015, www.bom.gov.au/climate/updates/articles/a010-southern-rainfall-decline.shtml.

91   **99 and 84 percent, respectively:** Albert I. J. M. van Dijk et al., "The Millennium Drought in Southeast Australia (2001–2009): Natural and Human Causes and Implications for Water Resources, Ecosystems, Economy, and Society," *Water Resources Research* 49 (February 2013): pp. 1040–57, http://doi.org/10.1002/wrcr.20123.

91   **wetlands turned acidic:** "Managing Water for the Environment During Drought: Lessons from Victoria, Australia, Technical Appendices," Public Policy Institute of California (San Francisco, June 2016), p. 8, www.ppic.org/content/pubs/other/0616JMR_appendix.pdf.

91   **for weeks in May and June:** Michael Safi, "Washing Is a Privilege: Life on the Frontline of India's Water Crisis," *The Guardian*, June 21, 2018. See also

Maria Abi-Habib and Hari Kumar, "Deadly Tensions Rise as India's Water Supply Runs Dangerously Low," *The New York Times*, June 17, 2018.

91   **United States west of Texas:** Mesfin M. Mekonnen and Arjen Y. Hoekstra, "Four Billion People Facing Severe Water Scarcity," *Science Advances* 2, no. 2 (February 2016), https://doi.org/10.1126/sciadv.1500323.

92   **water demand from the global food system:** World Bank, "High and Dry," p. 5.

92   **"the impacts of climate change":** Ibid., p. vi.

92   **regional GDP could decline:** Ibid., p. 13.

92   **list of all armed conflicts:** "Water Conflict," Pacific Institute: The World's Water, May 2018. www.worldwater.org/water-conflict.

93   **the number of cholera cases:** International Committee of the Red Cross, "Health Crisis in Yemen," www.icrc.org/en/where-we-work/middle-east/yemen/health-crisis-yemen.

Dying Oceans

94   **"Undersea":** Carson was just thirty when she published her essay in *The Atlantic,* still working as a biologist for the Fisheries Bureau of the U.S. Fish and Wildlife Service. In the oceans, she wrote, "we see parts of the plan fall into place: the water receiving from earth and air the simple materials, storing them up until the gathering energy of the spring sun wakens the sleeping plants to a burst of dynamic activity, hungry swarms of planktonic animals growing and multiplying upon the abundant plants, and themselves falling prey to the shoals of fish; all, in the end; to be redissolved into their component substances when the inexorable laws of the sea demand it. Individual elements are lost to view, only to repair again and again in different incarnations in a kind of material immortality. Kindred forces to those which, in some period inconceivably remote, gave birth to that primeval bit of protoplasm tossing on the ancient seas continue their mighty and incomprehensible work. Against this cosmic background the lifespan of a particular plant or animal appears, not as drama complete in itself, but only as a brief interlude in a panorama of endless change."

94   **70 percent of the earth's surface:** National Ocean Service, "How Much Water Is in the Ocean?" National Oceanic and Atmospheric Administration, June 25, 2018, https://oceanservice.noaa.gov/facts/oceanwater.html.

94   **seafood accounts for nearly a fifth:** "Availability and Consumption of Fish," World Health Organization, www.who.int/nutrition/topics/3_foodconsumption/en/index5.html.

95   **fish populations have migrated:** Malin L. Pinsky et al., "Preparing Ocean Governance for Species on the Move," *Science* 360, no. 6394 (June 2018): pp. 1189–91, https://doi.org/10.1126/science.aat2360.

95   **13 percent of the ocean undamaged:** Kendall R. Jones et al., "The Location and Protection Status of Earth's Diminishing Marine Wilderness," *Current Biology* 28, no. 15 (August 2018): pp. 2506–12, https://doi.org/10.1016/j.cub.2018.06.010.

95   **parts of the Arctic have been so transformed:** Sigrid Lind et al., "Arctic Warming Hotspot in the Northern Barents Sea Linked to Declining Sea-Ice Import," *Nature Climate Change* 8 (June 2018): pp. 634–39, https://doi.org/10.1038/s41558-018-0205-y.

95   **more than a fourth of the carbon emitted:** Rob Monroc, "How Much CO_2 Can the Oceans Take Up?" Scripps Institution of Oceanography, July 13, 2013, https://scripps.ucsd.edu/programs/keelingcurve/2013/07/03/how-much-co2-can-the-oceans-take-up.

95   **90 percent of global warming's excess heat:** Peter J. Gleckler et al., "Industrial-Era Global Ocean Heat Uptake Doubles in Recent Decades," *Nature Climate Change* 6 (January 2016): pp. 394–98, https://doi.org/10.1038/nclimate2915.

95   **absorbing three times as much:** Ibid.

96   **90 percent of the energy needs:** Australian Government Great Barrier Reef Marine Park Authority, "Managing the Reef."

96   **Great Barrier Reef:** Robinson Meyer, "Since 2016, Half of All Coral in the Great Barrier Reef Has Died," *The Atlantic*, April 2018.

96   **from 2014 to 2017:** Michon Scott and Rebecca Lindsey, "Unprecedented Three Years of Global Coral Bleaching, 2014–2017," Climate.gov, August 1, 2018, www.climate.gov/news-features/understanding-climate/unprecedented-3-years-global-coral-bleaching-2014%E2%80%932017.

96   **"twilight zone":** C. C. Baldwin et al., "Below the Mesophotic," *Scientific Reports* 8, no. 4920 (March 2018), https://doi.org/10.1038/s41598-018-23067-1.

96   **threaten 90 percent of all reefs:** Lauretta Burke et al., "Reefs at Risk Revisited," World Resources Institute (Washington, D.C., 2011), p. 6, https://wriorg.s3.amazonaws.com/s3fs-public/pdf/reefs_at_risk_revisited.pdf.

96   **as much as a quarter of all marine life:** Ocean Portal Team, "Corals and Coral Reefs," *Smithsonian*, April 2018, https://ocean.si.edu/ocean-life/invertebrates/corals-and-coral-reefs.

96   **food and income for half a billion:** "Coral Ecosystems," National Oceanic and Atmospheric Administration, www.noaa.gov/resource-collections/coral-ecosystems.

96   **worth at least $400 million:** Michael W. Beck et al., "The Global Flood Protection Savings Provided by Coral Reefs," *Nature Communications* 9, no. 2186 (June 2018), https://doi.org/10.1038/s41467-018-04568-z.

96   **oysters and mussels will struggle:** Kate Madin, "Ocean Acidification: A Risky Shell Game," *Oceanus Magazine*, December 4, 2009, www.whoi.edu/oceanus/feature/ocean-acidification--a-risky-shell-game.

96   **fishes' sense of smell:** Cosima Porteus et al., "Near-Future CO_2 Levels Impair the Olfactory System of Marine Fish," *Nature Climate Change* 8 (July 23, 2018).

96   **32 percent in just ten years:** Graham Edgar and Trevor J. Ward, "Australian Commercial Fish Populations Drop by a Third over Ten Years," *The Conversation*, June 6, 2018, https://theconversation.com/australian-commercial-fish-populations-drop-by-a-third-over-ten-years-97689.

96   **by a factor perhaps as large as a thousand:** Jurriaan M. De Vos et al., "Estimating the Normal Background Rate of Species Extinction," *Conservation Biology*, August 26, 2014.

97   **an era marked by ocean acidification:** A. H. Altieri and K. B. Gedan, "Climate Change and Dead Zones," *Global Change Biology* (November 10, 2014), https://doi.org/10.1111/gcb.12754.

97   **with no oxygen at all:** "SOS: Is Climate Change Suffocating Our Seas?" National Science Foundation, www.nsf.gov/news/special_reports/deadzones/climatechange.jsp.

97   **a dead zone the size of Florida:** Bastien Y. Queste et al., "Physical Controls on Oxygen Distribution and Denitrification Potential in the North West Arabian Sea," *Geophysical Research Letters* 45, no. 9 (May 2018). See also "Growing 'Dead Zone' Confirmed by Underwater Robots" (press release), University of East Anglia, April 27, 2018, www.uea.ac.uk/about/-/growing-dead-zone-confirmed-by-underwater-robots-in-the-gulf-of-oman.

97   **Dramatic declines in ocean oxygen:** Peter Brannen, "A Foreboding Similarity in Today's Oceans and a 94-Million-Year-Old Catastrophe," *The Atlantic*, January 12, 2018. See also Dana Nuccitelli, "Burning Coal May Have Caused Earth's Worst Mass Extinction," *The Guardian*, March 12, 2018.

98   **trip can take a thousand years:** National Ocean Service, "Currents: The Global Conveyor Belt," National Oceanic and Atmospheric Administration, https://oceanservice.noaa.gov/education/tutorial_currents/05conveyor2.html.

98   **depressed the velocity of the Gulf Stream:** Stefan Rahmstorf et al., "Exceptional Twentieth-Century Slowdown in Atlantic Ocean Overturning Circulation," *Nature Climate Change* 5 (May 2015), https://doi.org/10.1038/nclimate2554.

98   **"an unprecedented event":** Ibid.

98   **two major papers:** L. Caesar et al., "Observed Fingerprint of a Weakening Atlantic Ocean Overturning Circulation," *Nature* 556 (April 2018): pp. 191–96, https://doi.org/10.1038/s41586-018-0006-5; David J. R. Thornalley et al., "Anomalously weak Labrador Sea convection and Atlantic overturning during the past 150 years," *Nature* 556 (April 2018), pp. 227–30, https://doi.org/10.1038/s41586-018-0007-4.

99   **"tipping point":** Joseph Romm, "Dangerous Climate Tipping Point Is 'About a Century Ahead of Schedule' Warns Scientist," *Think Progress*, April 12, 2018.

Unbreathable Air

100   **cognitive ability declines:** Joseph Romm, *Climate Change: What Everyone Needs to Know* (New York: Oxford University Press, 2016), p. 113.

100   **almost a quarter of those surveyed in Texas:** Ibid., p. 114.

101   **deaths from dust pollution:** Ploy Achakulwisut et al., "Drought Sensitivity in Fine Dust in the U.S. Southwest," *Environmental Research Letters* 13 (May 2018), https://doi.org/10.1088/1748-9326/aabf20.

101   **a 70 percent increase:** G. G. Pfister et al., "Projections of Future Summertime Ozone over the U.S.," *Journal of Geophysical Research Atmospheres* 119, no. 9 (May 2014): pp. 5559–82, https://doi.org/10.1002/2013JD020932.

101   **2 billion people globally:** Romm, *Climate Change*, p. 105.

101   **10,000 people die:** DARA, *Climate Vulnerability Monitor: A Guide to the Cold Calculus of a Hot Planet*, 2nd ed. (Madrid, 2012), p. 17, https://daraint.org/wp-content/uploads/2012/10/CVM2-Low.pdf. James Hansen himself has made this comparison in a number of venues, including in an interview with me published in *New York* as "Climate Scientist James Hansen: 'The Planet Could Become Ungovernable,'" July 12, 2017.

101   **researchers call the effect "huge":** Xin Zhang et al., "The Impact of Exposure to Air Pollution on Cognitive Performance," *Proceedings of the National Academy of Sciences* 155, no. 37 (September 2018): pp. 9193–97, https://doi.org/10.1073/pnas.1809474115. Coauthor Xi Chen made the "huge" comment to a number of news outlets, including *The Guardian*: Damian Carrington and Lily Kuo, "Air Pollution Causes 'Huge' Reduction in Intelligence, Study Reveals," August 27, 2018.

102   **Simple temperature rise:** Joshua Goodman et al., "Heat and Learning" (National Bureau of Economic Research working paper no. 24639, May 2018), https://doi.org/10.3386/w24639.

102   **increased mental illness in children:** Anna Oudin et al., "Association Between Neighbourhood Air Pollution Concentrations and Dispensed Medication for Psychiatric Disorders in a Large Longitudinal Cohort of Swedish Children and Adolescents," *BMJ Open* 6, no. 6 (June 2016), https://doi.org/10.1136/bmjopen-2015-010004.

102   **likelihood of dementia in adults:** Hong Chen et al., "Living near Major Roads and the Incidence of Dementia, Parkinson's Disease, and Multiple Sclerosis: A Population-Based Cohort Study," *The Lancet* 389, no. 10070 (February 2017), pp. 718–26, https://doi.org/10.1016/S0140-6736(16)32399-6.

102   **reduce earnings and labor force participation:** Adam Isen et al., "Every Breath You Take—Every Dollar You'll Make: The Long-Term Consequences of the Clean Air Act of 1970" (National Bureau of Economic Research working paper no. 19858, September 2015), https://doi.org/10.3386/w19858.

102   **E-ZPass:** Janet Currie and W. Reed Walker, "Traffic Congestion and Infant Health: Evidence from E-ZPass" (National Bureau of Economic Research working paper no. 15413, April 2012), https://doi.org/10.3386/w15413.

102   **melting Arctic ice remodeled Asian weather patterns:** Yufei Zou et al., "Arctic Sea Ice, Eurasia Snow, and Extreme Winter Haze in China," *Science Advances* 3, no. 3 (March 2017), https://doi.org/10.1126/sciadv.1602751.

102   **peak Air Quality Index of 993:** Steve LeVine, "Pollution Score: Beijing 993, New York 19," *Quartz*, January 14, 2013, https://qz.com/43298/pollution-score-beijing-993-new-york-19.

102   **new and unstudied kind of smog:** Lijian Han et al., "Multicontaminant Air Pollution in Chinese Cities," *Bulletin of the World Health Organization* 96 (February 2018): pp. 233–42E, http://dx.doi.org/10.2471/BLT.17.195560; Fred Pearce, "How a 'Toxic Cocktail' Is Posing a Troubling Health Risk in China's Cities," *Yale Environment 360*, April 17, 2018, https://e360.yale.edu/features/how-a-toxic-cocktail-is-posing-a-troubling-health-risk-in-chinese-cities.

102   **1.37 million deaths:** Jun Liu et al., "Estimating Adult Mortality Attributable to PM2.5 Exposure in China with Assimilated PM2.5 Concentrations Based

on a Ground Monitoring Network," *Science of the Total Environment* 568 (October 2016): pp. 1253–62, https://doi.org/10.1016/j.scitotenv.2016.05.165.

103   **the air around San Francisco:** Michelle Robertson, "It's Not Just Fog Turning the Sky Gray: SF Air Quality Is Three Times Worse than Beijing," *SF Gate,* August 23, 2018.

103   **In Seattle:** In August 2018, the mayor's office tweeted, "Today's air quality has been declared UNHEALTHY FOR ALL GROUPS. Stay inside, limit outdoor work, and try not to drive."

103   **Air Quality Index reached 999:** Rachel Feltman, "Air Pollution in Delhi Is Literally off the Charts," *Popular Science*, November 8, 2016.

103   **more than two packs of cigarettes:** Richard A. Muller and Elizabeth A. Muller, "Air Pollution and Cigarette Equivalence," *Berkeley Earth*, http://berkeleyearth.org/air-pollution-and-cigarette-equivalence.

103   **patient surge of 20 percent:** Durgesh Nandan Jha, "Pollution Causing Arthritis to Flare Up, 20% Rise in Patients at Hospitals," *The Times of India,* November 11, 2017.

103   **cars crashed in pileups:** "Blinding Smog Causes 24-Vehicle Pile-up on Expressway near Delhi," *NDTV,* November 8, 2017.

103   **United canceled flights:** Catherine Ngai, Jamie Freed, and Henning Gloystein, "United Resumes Newark-Delhi Flights After Halt Due to Poor Air Quality," Reuters, November 12, 2017, https://www.reuters.com/article/us-airlines-india-pollution/united-resumes-newark-delhi-flights-after-halt-due-to-poor-air-quality-idUSKBN1DC142?il=0.

103   **even short-term exposure:** Benjamin D. Horne et al., "Short-Term Elevation of Fine Particulate Matter Air Pollution and Acute Lower Respiratory Infection," *American Journal of Respiratory and Critical Care Medicine* 198, no. 6, (September 2018), https://doi.org/10.1164/rccm.201709-1883OC.

103   **nine million premature deaths:** Pamela Das and Richard Horton, "Pollution, Health, and the Planet: Time for Decisive Action," *The Lancet* 391, no. 10119 (October 2017): pp. 407–8, https://doi.org/10.1016/S0140-6736(17)32588-6.

103   **prevalence of stroke:** Kuam Ken Lee et al., "Air Pollution and Stroke," *Journal of Stroke* 20, no. 1 (January 2018): pp. 2–11, https://doi.org/10.5853/jos.2017.02894.

103   **heart disease:** R. D. Brook et al., "Particulate Matter Air Pollution and Cardiovascular Disease: An Update to the Scientific Statement from the American Heart Association," *Circulation* 121, no. 21 (June 2010): pp. 2331–78, https://doi.org/10.1161/CIR.0b013e3181dbece1.

104   **cancer of all kinds:** Kate Kelland and Stephanie Nebehay, "Air Pollution a Leading Cause of Cancer—U.N. Agency," Reuters, October 17, 2013, www.reuters.com/article/us-cancer-pollution/air-pollution-a-leading-cause-of-cancer-u-n-agency-idUSBRE99G0BB20131017.

104   **acute and chronic respiratory diseases:** Michael Guarnieri and John R. Balmes, "Outdoor Air Pollution and Asthma," *The Lancet* 383, no. 9928 (May 2014), https://doi.org/10.1016/S0140-6736(14)60617-6.

104   **adverse pregnancy outcomes:** Jessica Glenza, "Millions of Premature Births Could Be Linked to Air Pollution, Study Finds," *The Guardian,* February 16, 2017.

104   **worse memory, attention, and vocabulary:** Nicole Wetsman, "Air Pollution Might Be the New Lead," *Popular Science,* April 5, 2018.

104   **ADHD:** Oddvar Myhre et al., "Early Life Exposure to Air Pollution Particulate Matter (PM) as Risk Factor for Attention Deficit/Hyperactivity Disorder (ADHD): Need for Novel Strategies for Mechanisms and Causalities," *Toxicology and Applied Pharmacology* 354 (September 2018): pp. 196–214, https://doi.org/10.1016/j.taap.2018.03.015.

104   **autism spectrum disorders:** Raanan Raz et al., "Autism Spectrum Disorder and Particulate Matter Air Pollution Before, During, and After Pregnancy: A Nested Case-Control Analysis Within the Nurses' Health Study II Cohort," *Environmental Health Perspectives* 123, no. 3 (March 2015): pp. 264–70, https://doi.org/10.1289/ehp.1408133.

104   **damage the development of neurons:** Sam Brockmeyer and Amedeo D'Angiulli, "How Air Pollution Alters Brain Development: The Role of Neuroinflammation," *Translational Neuroscience* 7 (March 2016): pp. 24–30, https://doi.org/10.1515/tnsci-2016-0005.

104   **deform your DNA:** Frederica Perera et al., "Shorter Telomere Length in Cord Blood Associated with Prenatal Air Pollution Exposure: Benefits of Intervention," *Environment International* 113 (April 2018): pp. 335–40, https://doi.org/10.1016/j.envint.2018.01.005.

104   **98 percent of cities:** World Health Organization, "WHO Global Urban Ambient Air Pollution Database," 2016, www.who.int/phe/health_topics/outdoorair/databases/cities/en.

104   **95 percent of the world's population:** Health Effects Institute, "State of Global Air 2018: A Special Report on Global Exposure to Air Pollution and Its Disease Burden" (Boston, 2018), p. 3, www.stateofglobalair.org/sites/default/files/soga-2018-report.pdf.

104   **more than a million Chinese each year:** Aaron J. Cohen et al., "Estimates and 25-Year Trends of the Global Burden of Disease Attributable to Ambient Air Pollution: An Analysis of Data from the Global Burden of Diseases Study 2015," *The Lancet* 389, no. 10082 (May 2017): pp. 1907–18, https://doi.org/10.1016/S0140-6736(17)30505-6.

104   **one out of six deaths:** Das and Horton, "Pollution, Health, and the Planet," https://doi.org/10.1016/S0140-6736(17)32588-6.

104   **"Great Pacific garbage patch":** *Smithsonian* calls it more of a "trash soup."

105   **700,000 of them can be released:** Imogen E. Napper and Richard C. Thompson, "Release of Synthetic Microplastic Fibres from Domestic Washing Machines: Effects of Fabric Type and Washing Conditions," *Marine Pollution Bulletin* 112, no. 1–2 (November 2016): pp. 39–45, http://dx.doi.org/10.1016/j.marpolbul.2016.09.025.

105   **a quarter of fish sold:** Kat Kerlin, "Plastic for Dinner: A Quarter of Fish Sold at Markets Contain Human-Made Debris," UC Davis, September 24,

2015, www.ucdavis.edu/news/plastic-dinner-quarter-fish-sold-markets
-contain-human-made-debris.

105   **11,000 bits each year:** Lisbeth Van Cauwenberghe and Colin R. Janssen, "Microplastics in Bivalves Cultured for Human Consumption," *Environmental Pollution* 193 (October 2014): pp. 65–70, https://doi.org/10.1016/j.envpol.2014.06.010.

105   **total number of marine species:** Clive Cookson, "The Problem with Plastic: Can Our Oceans Survive?" *Financial Times,* January 23, 2018.

105   **73 percent of fish surveyed:** Alina M. Wieczorek et al., "Frequency of Microplastics in Mesopelagic Fishes from the Northwest Atlantic," *Frontiers in Marine Science* (February 2018), https://doi.org/10.3389/fmars.2018.00039.

105   **every 100 grams of mussels:** Jiana Lee et al., "Microplastics in Mussels Sampled from Coastal Waters and Supermarkets in the United Kingdom," *Environmental Pollution* 241 (October 2018): pp. 35–44, https://doi.org/10.1016/j.envpol.2018.05.038.

105   **Some fish have learned to eat plastic:** Matthew S. Savoca et al., "Odours from Marine Plastic Debris Induce Food Search Behaviours in a Forage Fish," *Proceedings of the Royal Society B Biological Sciences* 284, no. 1860 (August 2017), https://doi.org/10.1098/rspb.2017.1000.

105   **bits that scientists are now calling "nanoplastics":** Amanda L. Dawson et al., "Turning Microplastics into Nanoplastics Through Digestive Fragmentation by Antarctic Krill," *Nature Communications* 9, no. 1001 (March 2018), https://doi.org/10.1038/s41467-018-03465-9.

105   **3.4 million microplastic particles:** Courtney Humphries, "Freshwater's Macro Microplastic Problem," *Nova*, May 11, 2017, www.pbs.org/wgbh/nova/article/freshwater-microplastics.

105   **225 pieces of plastic:** Cookson, "The Problem with Plastic."

105   **sixteen of seventeen tested brands:** Ali Karami et al., "The Presence of Microplastics in Commercial Salts from Different Countries," *Scientific Reports* 7, no. 46173 (April 2017), https://doi.org/10.1038/srep46173.

105   **one million times more toxic:** 5 Gyres: Science to Solutions, "Take Action: Microbeads," www.5gyres.org/microbeads.

106   **We can breathe in microplastics:** Johnny Gasperi et al., "Microplastics in Air: Are We Breathing It In?" *Current Opinion in Environmental Science and Health* 1 (February 2018): pp. 1–5, https://doi.org/10.1016/j.coesh.2017.10.002.

106   **94 percent of all tested American cities:** Dan Morrison and Christopher Tyree, "Invisibles: The Plastic Inside Us," *Orb* (2017), https://orbmedia.org/stories/Invisibles_plastics.

106   **expected to triple by 2050:** World Economic Forum, *The New Plastics Economy: Rethinking the Future of Plastics* (Cologny, Switz.: January 2016), p. 10.

106   **they release methane and ethylene:** Sarah-Jeanne Royer et al., "Production of Methane and Ethylene from Plastic in the Environment," *PLOS One* 13, no. 8 (August 2018), https://doi.org/10.1371/journal.pone.0200574.

106   **Aerosol products actually suppress:** B. H. Samset et al., "Climate Impacts from a Removal of Anthropogenic Aerosol Emissions," *Geophysi-*

cal Research Letters 45, no. 2 (January 2018): pp. 1020–29, https://doi .org/10.1002/2017GL076079.

106   **only heated up two-thirds as much:** Samset, "Climate Impacts from a Removal," https://doi.org/10.1002/2017GL076079. Samset himself says, "Global warming to date is one degree Celsius (or thereabouts). Our paper showed that industrial/human induced aerosol emissions mask about half a degree of additional warming." And because of how unevenly warming is distributed across the planet, he adds, "we note that in two models, Arctic warming due to aerosol reductions reaches 4°C in some locations."

107   **"Catch-22":** P. J. Crutzen, "Albedo Enhancement by Stratospheric Sulfur Injections: A Contribution to Resolve a Policy Dilemma?" *Climatic Change* 77 (2006): pp. 211–19, https://doi.org/ 10.1007/s10584-006-9101-y.

107   **"devil's bargain":** Eric Holthaus, "Devil's Bargain," *Grist*, February 8, 2018, https://grist.org/article/geoengineering-climate-change-air-pollution -save-planet.

107   **millions of lives each year:** This estimate of deaths from air pollution comes from the World Health Organization.

107   **tens of thousands of additional premature deaths:** Sebastian D. Eastham et al., "Quantifying the Impact of Sulfate Geoengineering on Mortality from Air Quality and UV-B Exposure," *Atmospheric Environment* 187 (August 2018): pp. 424–34, https://doi.org/10.1016/j.atmosenv.2018.05.047.

108   **rapidly dry the Amazon:** Christopher H. Trisos et al., "Potentially Dangerous Consequences for Biodiversity of Solar Geoengineering Implementation and Termination," *Nature Ecology and Evolution* 2 (January 2018), pp. 472–82, https://doi.org/10.1038/s41559-017-0431-0.

108   **negative effect on plant growth:** Jonathan Proctor et al., "Estimating Global Agricultural Effects of Geoengineering Using Volcanic Eruptions," *Nature* 560 (August 2018): pp. 480–83, https://doi.org/10.1038/s41586-018-0417-3.

Plagues of Warming

109   **diseases that have not circulated:** Jasmin Fox-Skelly, "There Are Diseases Hidden in Ice, and They Are Waking Up," BBC, May 4, 2017, www.bbc .com/earth/story/20170504-there-are-diseases-hidden-in-ice-and-they-are -waking-up.

109   **"extremophile" bacteria:** "NASA Finds Life at 'Extremes,'" NASA, February 24, 2005, www.nasa.gov/vision/earth/livingthings/extremophile1.html.

109   **an 8-million-year-old bug:** Kay D. Bidle et al., "Fossil Genes and Microbes in the Oldest Ice on Earth," *Proceedings of the National Academies of Science* 104, no. 33 (August 2007): pp. 13455–60, https://doi.org/10.1073/ pnas.0702196104.

109   **a Russian scientist self-injected:** Jordan Pearson, "Meet the Scientist Who Injected Himself with 3.5 Million-Year-Old Bacteria," *Motherboard*, December 9, 2015, https://motherboard.vice.com/en_us/article/yp3gg7/ meet-the-scientist-who-injected-himself-with-35-million-year-old-bacteria.

109   **a worm that had been frozen:** Mike McRae, "A Tiny Worm Frozen in Siberian Permafrost for 42,000 Years Was Just Brought Back to Life," *Science*

Alert, July 27, 2018, www.sciencealert.com/40-000-year-old-nematodes
-revived-siberian-permafrost.

110   **remnants of the 1918 flu:** Jeffery K. Taubenberger et al., "Discovery and
Characterization of the 1918 Pandemic Influenza Virus in Historical Context," *Antiviral Therapy* 12 (2007): pp. 581–91.

110   **infected as many as 500 million and killed as many as 50 million:** Centers
for Disease Control and Prevention, "Remembering the 1918 Influenza Pandemic," www.cdc.gov/features/1918-flu-pandemic/index.html; Jeffery K.
Taubenberger and David Morens, "1918 Influenza: The Mother of All Pandemics," *Emerging Infectious Diseases* 12, no.1 (January 2006): pp. 15–22,
https://dx.doi.org/10.3201/eid1201.050979.

110   **3 percent of the world's population:** U.S. Census Bureau, "Historical Estimates of World Population," www.census.gov/data/tables/time-series/demo/
international-programs/historical-est-worldpop.html.

110   **smallpox:** "Experts Warn of Threat of Born-Again Smallpox from Old Siberian Graveyards," *The Siberian Times*, August 12, 2016, https://siberiantimes
.com/science/opinion/features/f0249-experts-warn-of-threat-of-born-again
-smallpox-from-old-siberian-graveyards.

110   **bubonic plague:** Fox-Skelly, "There Are Diseases Hidden in Ice."

110   **among many other diseases:** Robinson Meyer, "The Zombie Diseases of Climate Change," *The Atlantic*, November 6, 2017.

110   **But in 2016, a boy:** Michaeleen Doucleff, "Anthrax Outbreak in Russia Thought to Be Result of Thawing Permafrost," NPR, August 3, 2016,
www.npr.org/sections/goatsandsoda/2016/08/03/488400947/anthrax
-outbreak-in-russia-thought-to-be-result-of-thawing-permafrost.

111   *Haemagogus* and *Sabethes* **mosquitoes:** World Health Organization, "Yellow Fever—Brazil," March 9, 2018, www.who.int/csr/don/09-march-2018
-yellow-fever-brazil.

111   **more than thirty million people:** Ibid.

111   **kills between 3 and 8 percent:** Shasta Darlington and Donald G. McNeil Jr.,
"Yellow Fever Circles Brazil's Huge Cities," *The New York Times*, March 8,
2018.

111   **Malaria alone kills:** World Health Organization, "Number of Malaria
Deaths," www.who.int/gho/malaria/epidemic/deaths. See also Centers for
Disease Control and Prevention, "Epidemiology," www.cdc.gov/dengue/
epidemiology/index.html.

111   **disease mutation:** "Zika Microcephaly Linked to Single Mutation," *Nature*,
October 3, 2017, www.nature.com/articles/d41586-017-04093-x.

111   **appear to cause birth defects:** Ling Yuan et al., "A Single Mutation in the
prM Protein of Zika Virus Contributes to Fetal Microcephaly," *Science* 358,
no. 6365 (November 2017): pp. 933–36, https://doi.org/10.1126/science
.aam7120.

112   **when another disease is present:** Declan Butler, "Brazil Asks Whether Zika
Acts Alone to Cause Birth Defects," *Nature*, July 25, 2016, www.nature.com/
news/brazil-asks-whether-zika-acts-alone-to-cause-birth-defects-1.20309.

112   **World Bank estimates that by 2030:** World Bank Group's Climate Change and Development Series, "Shock Waves: Managing the Impacts of Climate Change on Poverty" (Washington, D.C., 2016), p. 119, https://openknowledge.worldbank.org/bitstream/handle/10986/22787/9781464806735.pdf.

112   **Lyme case counts have spiked:** Mary Beth Pfeiffer, *Lyme: The First Epidemic of Climate Change* (Washington, D.C.: Island Press, 2018), pp. 3–13.

112   **300,000 new infections each year:** Centers for Disease Control and Prevention, "Lyme and Other Tickborne Diseases," www.cdc.gov/media/dpk/diseases-and-conditions/lyme-disease/index.html.

112   **fleas have tripled in the U.S.:** Centers for Disease Control and Prevention, "Illnesses from Mosquito, Tick, and Flea Bites Increasing in the U.S.," May 1, 2018, www.cdc.gov/media/releases/2018/p0501-vs-vector-borne.html.

112   **encountering ticks for the first time:** Avichai Scher and Lauren Dunn, "'Citizen Scientists' Take On Growing Threat of Tick-Borne Diseases," NBC News, July 12, 2018, www.nbcnews.com/health/health-news/citizen-scientists-take-growing-threat-tick-borne-diseases-n890996.

112   **winter ticks helped drop the moose population:** Center for Biological Diversity, "Saving the Midwestern Moose," www.biologicaldiversity.org/species/mammals/midwestern_moose/index.html.

113   **90,000 engorged ticks:** Katie Burton, "Climate-Change Triggered Ticks Causing Rise in 'Ghost Moose,'" *Geographical*, November 27, 2018, http://geographical.co.uk/nature/wildlife/item/3008-ghost-moose.

113   **a million yet-to-be-discovered viruses:** Dennis Carroll et al., "The Global Virome Project," *Science* 359, no. 6378 (February 2018): pp. 872–74, https://doi.org/10.1126/science.aap7463.

113   **More than 99 percent:** Nathan Collins, "Stanford Study Indicates That More than 99 Percent of the Microbes Inside Us Are Unknown to Science," *Stanford News*, August 22, 2017, https://news.stanford.edu/2017/08/22/nearly-microbes-inside-us-unknown-science.

114   **the case of the saiga:** Ed Yong, "Why Did Two-Thirds of These Weird Antelope Suddenly Drop Dead?" *The Atlantic*, January 17, 2018.

114   **nearly two-thirds of the global population:** Richard A. Kock et al., "Saigas on the Brink: Multidisciplinary Analysis of the Factors Influencing Mass Mortality Events," *Science Advances* 4, no. 1 (January 2018), https://doi.org/10.1126/sciadv.aao2314.

Economic Collapse

116   **"Whoever says Industrial Revolution":** Eric Hobsbawm, *Industry and Empire: The Birth of the Industrial Revolution* (New York: The New Press, 1999), p. 34.

117   **about one percentage point:** Solomon Hsiang et al., "Estimating Economic Damage from Climate Change in the United States," *Science* 356, no. 6345 (June 2017): 1362–69, https://doi.org/10.1126/science.aal4369.

117   **23 percent loss in per capita:** Marshall Burke et al., "Global Non-Linear Effect of Temperature on Economic Production," *Nature* 527 (October 2015): pp. 235–39, https://doi.org/10.1038/nature15725.

117   **There is a 51 percent chance:** Marshall Burke, "Economic Impact of Climate Change on the World," http://web.stanford.edu/~mburke/climate/map.php.

117   **a team led by Thomas Stoerk:** Thomas Stoerk et al., "Recommendations for Improving the Treatment of Risk and Uncertainty in Economic Estimates of Climate Impacts in the Sixth Intergovernmental Panel on Climate Change Assessment Report," *Review of Environmental Economics and Policy* 12, no. 2 (August 2018): pp. 371–76, https://doi.org/10.1093/reep/rey005.

118   **global boom of the early 1960s:** World Bank, "GDP Growth (Annual %)," https://data.worldbank.org/indicator/NY.GDP.MKTP.KD.ZG.

118   **There are places that benefit:** Burke, "Economic Impact of Climate Change," http://web.stanford.edu/~mburke/climate/map.php.

118   **India alone, one study proposed:** Katharine Ricke et al., "Country-Level Social Cost of Carbon," *Nature Climate Change* 8 (September 2018): pp. 895––900, http://doi.org/10.1038/s41558-018-0282-y.

118   **800 million:** World Bank, "South Asia's Hotspots: Impacts of Temperature and Precipitation Changes on Living Standards" (Washington, D.C., 2018), p. xi.

118   **dragged into extreme poverty:** World Bank Group's Climate Change and Development Series, "Shock Waves: Managing the Impacts of Climate Change on Poverty" (Washington, D.C., 2016), p. xi, https://openknowledge.worldbank.org/bitstream/handle/10986/22787/9781464806735.pdf.

119   **chronic flooding by 2100:** Union of Concerned Scientists, "Underwater: Rising Seas, Chronic Floods, and the Implications for U.S. Coastal Real Estate" (Cambridge, MA, 2018), p. 5, www.ucsusa.org/global-warming/global-warming-impacts/sea-level-rise-chronic-floods-and-us-coastal-real-estate-implications.

119   **$30 billion in New Jersey:** Union of Concerned Scientists, "New Study Finds 251,000 New Jersey Homes Worth $107 Billion Will Be at Risk from Tidal Flooding," June 18, 2018, www.ucsusa.org/press/2018/new-study-finds-251000-new-jersey-homes-worth-107-billion-will-be-risk-tidal-flooding#.W-o1FehKg2x.

120   **which is now commonplace:** Zach Wichter, "Too Hot to Fly? Climate Change May Take a Toll on Flying," *The New York Times*, June 20, 2017.

120   **Every round-trip plane ticket:** Dirk Notz and Julienne Stroeve, "Observed Arctic Sea-Ice Loss Directly Follows Anthropogenic CO_2 Emission," *Science* 354, no. 6313 (November 2016): pp. 747–50, https://doi.org/10.1126/science.aag2345.

120   **From Switzerland to Finland:** Olav Vilnes et al., "From Finland to Switzerland—Firms Cut Output Amid Heatwave," *Montel News*, July 27, 2018, www.montelnews.com/en/story/from-finland-to-switzerland--firms-cut-output-amid-heatwave/921390.

120   **670 million lost power:** Jim Yardley and Gardiner Harris, "Second Day of Power Failures Cripples Wide Swath of India," *The New York Times,* July 31, 2012.

121   **13 degrees Celsius:** Burke, "Global Non-Linear Effect of Temperature," https://doi.org/10.1038/nature15725; author interview with Marshall Burke.

121   **Already-hot countries:** World Bank, "South Asia's Hotspots."

121   **up to 20 percent:** Hsiang, "Estimating Economic Damage from Climate Change," https://doi.org/10.1126/science.aal4369.

121   **"economic ripple effect":** Zhengtao Zhang et al., "Analysis of the Economic Ripple Effect of the United States on the World Due to Future Climate Change," *Earth's Future* 6, no. 6 (June 2018): pp. 828–40, https://doi .org/10.1029/2018EF000839.

122   **negative $26 trillion:** The New Climate Economy, "Unlocking the Inclusive Growth Story of the 21st Century: Accelerating Climate Action in Urgent Times" (Washington, D.C.: Global Commission on the Economy and Climate, September 2018), p. 8, https://newclimateeconomy.report/2018.

123   **growth consequences of some scenarios:** Marshall Burke et al., "Large Potential Reduction in Economic Damages Under U.N. Mitigation Targets," *Nature* 557 (May 2018): pp. 549–53, https://doi.org/10.1038/ s41586-018-0071-9.

Climate Conflict

124   **for every half degree of warming:** Solomon M. Hsiang et al., "Quantifying the Influence of Climate on Human Conflict," *Science* 341, no. 6151 (September 2013), https://doi.org/10.1126/science.1235367.

125   **elevated Africa's risk of conflict:** Tamma A. Carleton and Solomon M. Hsiang, "Social and Economic Impacts of Climate," *Science* 353, no. 6304 (September 2016), http://doi.org/10.1126/science.aad9837.

125   **393,000 additional deaths:** Marshall B. Burke et al., "Warming Increases the Risk of Civil War in Africa," *Proceedings of the National Academy of Sciences* 106, no. 49 (December 2009): pp. 20670–74, https://doi.org/10.1073/ pnas.0907998106. This would represent a 54 percent increase.

125   **The drowning of American navy bases:** Union of Concerned Scientists, "The U.S. Military on the Front Lines of Rising Seas" (Cambridge, MA, 2016), www.ucsusa.org/global-warming/science-and-impacts/impacts/sea-level -rise-flooding-us-military-bases#.W-pKUuhKg2x.

126   **its islands will be underwater:** "We show that, on the basis of current greenhouse-gas emission rates, the nonlinear interactions between sea-level rise and wave dynamics over reefs will lead to the annual wave-driven overwash of most atoll islands by the mid-21st century. This annual flooding will result in the islands becoming uninhabitable because of frequent damage to infrastructure and the inability of their freshwater aquifers to recover between overwash events." Curt D. Storlazzi et al., "Most Atolls Will Be Uninhabitable by the Mid-21st Century Because of Sea-Level Rise Exacerbating

Wave-Driven Flooding," *Science Advances* 4, no. 4 (April 2018), https://doi .org/10.1126/sciadv.aap9741.

126 **the world's largest nuclear waste site:** Kim Wall, Coleen Jose, and Jan Henrik Hinzel, "The Poison and the Tomb: One Family's Journey to Their Contaminated Home," *Mashable*, February 25, 2018.

126 **From Boko Haram to ISIS:** Katharina Nett and Lukas Rüttinger, "Insurgency, Terrorism and Organised Crime in a Warming Climate: Analysing the Links Between Climate Change and Non-State Armed Groups," Climate Diplomacy (Berlin: Adelphi, October 2016).

126 **23 percent of conflict:** Carl-Friedrich Schleussner et al., "Armed-Conflict Risks Enhanced by Climate-Related Disasters in Ethnically Fractionalized Countries," *Proceedings of the National Academy of Sciences* 113, no. 33 (August 2016): pp. 9216–21, https://doi.org/10.1073/pnas.1601611113.

127 **"extreme risk":** Verisk Maplecroft, "Climate Change and Environmental Risk Atlas 2015" (Bath, UK, October 2014), www.maplecroft.com/portfolio/ new-analysis/2014/10/29/climate-change-and-lack-food-security-multiply -risks-conflict-and-civil-unrest-32-countries-maplecroft.

127 **What accounts for the relationship:** Christian Parenti, *Tropic of Chaos: Climate Change and the New Geography of Violence* (New York: Nation Books, 2011).

127 **the forced migration that can result:** Rafael Reuveny, "Climate Change– Induced Migration and Violent Conflict," *Political Geography* 26, no. 6 (August 2007): pp. 656–73, https://doi.org/10.1016/j.polgeo.2007.05.001.

127 **seventy million displaced:** Adrian Edwards, "Forced Displacement at Record 68.5 Million," UNHCR: The U.N. Refugee Agency, June 19, 2018, www.unhcr.org/en-us/news/stories/2018/6/5b222c494/forced-displacement -record-685-million.html.

127 **Egypt, Akkadia, Rome:** William Wan, "Ancient Egypt's Rulers Mishandled Climate Disasters. Then the People Revolted," *The Washington Post*, October 17, 2017; H. M. Cullen et al., "Climate Change and the Collapse of the Akkadian Empire: Evidence from the Deep Sea," *Geology* 28, no. 4 (April 2000): pp. 379–82; Kyle Harper, "How Climate Change and Disease Helped the Fall of Rome," *Aeon*, December 15, 2017, https://aeon.co/ideas/ how-climate-change-and-disease-helped-the-fall-of-rome.

127 **six categories:** Center for Climate and Security, "Epicenters of Climate and Security: The New Geostrategic Landscape of the Anthropocene" (Washington, D.C., June 2017), pp. 12–17, https://climateandsecurity.files .wordpress.com/2017/06/1_eroding-sovereignty.pdf.

128 **linguist Steven Pinker:** For Pinker's case for the world's improvement, see *Better Angels of Our Nature: Why Violence Has Declined* (New York: Viking, 2012); for his argument about why we can't appreciate that improvement, see *Enlightenment Now: The Case for Reason, Science, Humanism, and Progress* (New York: Viking, 2018).

129 **increases violent crime rates:** Leah H. Schinasi and Ghassan B. Hamra, "A Time Series Analysis of Associations Between Daily Temperature and Crime

Events in Philadelphia, Pennsylvania," *Journal of Urban Health* 94, no. 6 (December 2017): pp. 892–900, http://dx.doi.org/10.1007/s11524-017-0181-y.

129   **swearing on social media:** Patrick Baylis, "Temperature and Temperament: Evidence from a Billion Tweets" (Energy Institute at Haas working paper, November 2015), https://ei.haas.berkeley.edu/research/papers/WP265.pdf.

129   **a major league pitcher:** Richard P. Larrick et al., "Temper, Temperature, and Temptation," *Psychological Sciences* 22, no. 4 (February 2011): pp. 423–28, http://dx.doi.org/10.1177/0956797611399292.

129   **the longer drivers will honk:** Douglas T. Kenrick et al., "Ambient Temperature and Horn Honking: A Field Study of the Heat/Aggression Relationship," *Environment and Behavior* (March 1986), https://doi.org/10.1177/0013916586182002.

129   **police officers are more likely to fire:** Aldert Vrij et al., "Aggression of Police Officers as a Function of Temperature: An Experiment with the Fire Arms Training System," *Journal of Community and Applied Social Psychology* 4, no. 5 (December 1994): pp. 365–70, https://doi.org/10.1002/casp.2450040505.

129   **an additional 22,000 murders:** Matthew Ranson, "Crime, Weather, and Climate Change," *Journal of Environmental Economics and Management* 67, no. 3 (May 2014): pp. 274–302, https://doi.org/10.1016/j.jeem.2013.11.008.

129   **every single crime category:** Jackson G. Lu et al., "Polluted Morality: Air Pollution Predicts Criminal Activity and Unethical Behavior," *Psychological Science* 29, no. 3 (February 2018): pp. 340–55, https://doi.org/10.1177/0956797617735807.

130   **"food insecure":** Nett and Rüttinger, "Insurgency, Terrorism and Organised Crime," p. 37.

130   **organized crime . . . exploded:** Ibid., p. 39.

130   **Sicilian mafia was produced by drought:** Daron Acemoglu, Giuseppe De Feo, and Giacomo De Luca, "Weak States: Causes and Consequences of the Sicilian Mafia," VOX CEPR Policy Portal, March 2, 2018, https://voxeu.org/article/causes-and-consequences-sicilian-mafia.

130   **fifth-highest homicide rate:** Nett and Rüttinger, "Insurgency, Terrorism and Organised Crime," p. 35.

130   **second most dangerous country in the world for children:** UNICEF, *Hidden in Plain Sight: A Statistical Analysis of Violence Against Children* (New York: United Nations Children's Fund, 2014), p. 35, http://files.unicef.org/publications/files/Hidden_in_plain_sight_statistical_analysis_EN_3_Sept_2014.pdf.

130   **could make both of them ungrowable:** Pablo Imbach et al., "Coupling of Pollination Services and Coffee Suitability from Climate Change," *Proceedings of the National Academy of Sciences* 114, no. 39 (September 2017): pp. 10438–42, https://doi.org/10.1073/pnas.1617940114; Martina K. Linnenluecke et al., "Implications of Climate Change for the Sugarcane Industry," WIREs Climate Change 9, no. 1 (January–February 2018), https://doi.org/10.1002/wcc.498.

"Systems"

131   **22 million of them:** "In Photos: Climate Change, Disasters and Displacement," UNHCR: The U.N. Refugee Agency, January 1, 2015, www.unhcr.org/en-us/climate-change-and-disasters.html.

131   **60,000 climate migrants:** Emily Schmall and Frank Bajak, "FEMA Sees Trailers Only as Last Resort After Harvey, Irma," Associated Press, September 10, 2017, https://apnews.com/7716fb84835b48808839fbc888e96fb7.

131   **the evacuation of nearly 7 million:** Greg Allen, "Lessons from Hurricane Irma: When to Evacuate and When to Shelter in Place," NPR, June 1, 2018, www.npr.org/2018/06/01/615293318/lessons-from-hurricane-irma-when-to-evacuate-and-when-to-shelter-in-place.

131   **13 million Americans:** Andrew D. King and Luke J. Harrington, "The Inequality of Climate Change from 1.5 to 2°C of Global Warming," *Geophysical Research Letters* 45, no. 10 (May 2018): pp. 5030–33, https://doi.org/10.1029/2018GL078430.

132   **greatest in the world's least developed:** Ibid.

132   **In 2011, a single heat wave:** Katinka X. Ruthrof et al., "Subcontinental Heat Wave Triggers Terrestrial and Marine, Multi-Taxa Responses," *Scientific Reports* 8 (August 2018): p. 13094, https://doi.org/10.1038/s41598-018-31236-5.

133   **"current and existential national security risk":** Parliament of Australia, "Implications of Climate Change for Australia's National Security, Final Report, Chapter 2," www.aph.gov.au/Parliamentary_Business/Committees/Senate/Foreign_Affairs_Defence_and_Trade/Nationalsecurity/Final%20Report/c02; Ben Doherty, "Climate Change an 'Existential Security Risk' to Australia, Senate Inquiry Says." *The Guardian*, May 17, 2018.

133   **More than 140 million:** World Bank, *Groundswell: Preparing for Internal Climate Migration* (Washington, D.C., 2018), p. xix, https://openknowledge.worldbank.org/handle/10986/29461.

133   **as many as a billion migrants:** International Organization for Migration, "Migration, Environment and Climate Change: Assessing the Evidence," United Nations (Geneva, 2009), p. 43.

134   **more than two-thirds of outbreaks:** Frank C. Curriero et al., "The Association Between Extreme Precipitation and Waterborne Disease Outbreaks in the United States, 1948–1994," *American Journal of Public Health* 91, no. 8 (August 2001), https://doi.org/10.2105/AJPH.91.8.1194.

134   **more than 400,000 in Milwaukee:** William R. Mac Kenzie et al., "A Massive Outbreak in Milwaukee of Cryptosporidium Infection Transmitted Through the Public Water Supply," *The New England Journal of Medicine* 331 (July 1994): pp. 161–67, https://doi.org/10.1056/NEJM199407213310304.

134   **in Vietnam, those who passed:** Thuan Q. Thai and Evangelos M. Falaris, "Child Schooling, Child Health, and Rainfall Shocks: Evidence from Rural Vietnam" (Max Planck Institute working paper, September 2011), www.demogr.mpg.de/papers/working/wp-2011-011.pdf.

134   **In India, the same cycle-of-poverty pattern:** Santosh Kumar, Ramona Molitor, and Sebastian Vollmer, "Children of Drought: Rainfall Shocks and Early

Child Health in Rural India" (working paper, 2014); Santosh Kumar and Sebastian Vollmer, "Drought and Early Childhood Health in Rural India," *Population and Development Review* (2016).

134   **diminishing cognitive ability:** R. K. Phalkey et al., "Systematic Review of Current Efforts to Quantify the Impacts of Climate Change on Undernutrition," *Proceedings of the National Academy of Sciences* 112, no. 33 (August 2015): pp. E4522–29, https://doi.org/10.1073/pnas.1409769112; Charmian M. Bennett and Sharon Friel, "Impacts of Climate Change on Inequities in Child Health," *Children* 1, no. 3 (December 2014): pp. 461–73, https://doi.org/10.3390/children1030461; Iffat Ghani et al., "Climate Change and Its Impact on Nutritional Status and Health of Children," *British Journal of Applied Science and Technology* 21, no. 2 (2017): pp. 1–15, https://doi.org/10.9734/BJAST/2017/33276; Kristina Reinhardt and Jessica Fanzo, "Addressing Chronic Malnutrition Through Multi-Sectoral, Sustainable Approaches," *Frontiers in Nutrition* 1, no. 13 (August 2014), https://doi.org/10.3389/fnut.2014.00013.

134   **In Ecuador, climate damage:** Ram Fishman et al., "Long-Term Impacts of High Temperatures on Economic Productivity" (George Washington University Institute for International Economic Policy working paper, October 2015), https://econpapers.repec.org/paper/gwiwpaper/2015-18.htm.

134   **measurable declines:** Adam Isen et al., "Relationship Between Season of Birth, Temperature Exposure, and Later Life Well-Being," *Proceedings of the National Academy of Sciences* 114, no. 51 (December 2017): pp. 13447–52, https://doi.org/10.1073/pnas.1702436114.

134   **An enormous study in Taiwan:** C. R. Jung et al., "Ozone, Particulate Matter, and Newly-Diagnosed Alzheimer's Disease," *Journal of Alzheimer's Disease* 44, no. 2 (2015): pp. 573–84, https://doi.org/10.3233/JAD-140855.

134   **Similar patterns:** Emily Underwood, "The Polluted Brain," *Science* 355, no. 6323 (January 2017): pp. 342–45, https://doi.org/10.1126/science.355.6323.342.

135   **"Want to fight climate change?":** Damian Carrington, "Want to Fight Climate Change? Have Fewer Children," *The Guardian*, July 12, 2017.

135   **"Add this to the list of decisions":** Maggie Astor, "No Children Because of Climate Change? Some People Are Considering It," *The New York Times*, February 5, 2018.

136   **a half of all those exposed:** Janna Trombley et al., "Climate Change and Mental Health," *American Journal of Nursing* 117, no. 4 (April 2017): pp. 44–52, https://doi.org/ 10.1097/01.NAJ.0000515232.51795.fa.

136   **In England, flooding:** M. Reacher et al., "Health Impacts of Flooding in Lewes," *Communicable Disease and Public Health* 7, no. 1 (March 2004): pp. 39–46.

136   **aftermath of Hurricane Katrina:** Mary Alice Mills et al., "Trauma and Stress Response Among Hurricane Katrina Evacuees," *American Journal of Public Health* 97 (April 2007): pp. S116-23, https://doi.org/10.2105/AJPH.2006.086678.

136   **Wildfires, curiously:** Grant N. Marshall et al., "Psychiatric Disorders Among Adults Seeking Emergency Disaster Assistance After a Wildland-

Urban Interface Fire," *Psychiatric Services* 58, no. 4 (April 2007): pp. 509–14, https://doi.org/10.1176/ps.2007.58.4.509.

136   **"I don't know of a single scientist":** Kevin J. Doyle and Lise Van Susteren, *The Psychological Effects of Global Warming on the United States: And Why the U.S. Mental Health Care System Is Not Adequately Prepared* (Merrifield, VA: National Wildlife Federation, 2012), p. 19, www.nwf.org/~/media/PDFs/ Global-Warming/Reports/Psych_Effects_Climate_Change_Full_3_23 .ashx.

136   **"climate depression":** Madeleine Thomas, "Climate Depression Is Real, Just Ask a Scientist," *Grist*, October 28, 2014, https://grist.org/climate-energy/ climate-depression-is-for-real-just-ask-a-scientist.

136   **"environmental grief":** Jordan Rosenfeld, "Facing Down 'Environmental Grief,'" *Scientific American*, July 21, 2016.

137   **Hurricane Andrew hit Florida:** Ernesto Caffo and Carlotta Belaise, "Violence and Trauma: Evidence-Based Assessment and Intervention in Children and Adolescents: A Systematic Review," in *The Mental Health of Children and Adolescents: An Area of Global Neglect*, ed. Helmut Rehmschmidt et al. (West Sussex, Eng.: Wiley, 2007), p. 141.

137   **soldiers returning from war:** "PTSD: A Growing Epidemic," *NIH Medline-Plus* 4, no. 1 (2009): pp. 10–14, https://medlineplus.gov/magazine/issues/ winter09/articles/winter09pg10-14.html.

137   **One especially detailed study:** Armen K. Goenjian et al., "Posttraumatic Stress and Depressive Reactions Among Nicaraguan Adolescents After Hurricane Mitch," *American Journal of Psychiatry* 158, no. 5 (May 2001): pp. 788–94, https://doi.org/10.1176/appi.ajp.158.5.788.

137   **both the onset and the severity:** Haris Majeed and Jonathan Lee, "The Impact of Climate Change on Youth Depression and Mental Health," *The Lancet* 1, no. 3 (June 2017): pp.E94–95, https://doi.org/10.1016/ S2542-5196(17)30045-1.

137   **Rising temperature and humidity:** S. Vida, "Relationship Between Ambient Temperature and Humidity and Visits to Mental Health Emergency Departments in Quebec," *Psychiatric Services* 63, no. 11 (November 2012): pp. 1150–53, https://doi.org/10.1176/appi.ps.201100485.

137   **spikes in proper inpatient admissions:** Alana Hansen et al., "The Effect of Heat Waves on Mental Health in a Temperate Australian City," *Environmental Health Perspectives* 116, no. 10 (October 2008): pp. 1369–75, https:// doi.org/10.1289/ehp.11339.

137   **Schizophrenics, especially:** Roni Shiloh et al., "A Significant Correlation Between Ward Temperature and the Severity of Symptoms in Schizophrenia Inpatients: A Longitudinal Study," *European Neuropsychopharmacology* 17, no. 6–7 (May–June 2007): pp. 478–82, https://doi.org/10.1016/j .euroneuro.2006.12.001.

138   **mood disorders, anxiety disorders:** Hansen, "The Effect of Heat Waves on Mental Health," https://doi.org/10.1289/ehp.11339.

138   **Each increase of a single degree:** Marshall Burke et al., "Higher Temperatures Increase Suicide Rates in the United States and Mexico," *Na-*

ture Climate Change 8 (July 2018): pp. 723–29, https://doi.org/10.1038/s41558-018-0222-x.

138   **59,000 suicides:** Tamma Carleton, "Crop-Damaging Temperatures Increase Suicide Rates in India," *Proceedings of the National Academy of the Sciences* 114, no. 33 (August 2017): pp. 8746–51, https://doi.org/10.1073/pnas.1701354114.

III.   The Climate Kaleidoscope

Storytelling

143   **On-screen, climate devastation:** One good academic survey of this phenomenon is E. Ann Kaplan, *Climate Trauma: Foreseeing the Future in Dystopian Film and Fiction* (New Brunswick, NJ: Rutgers University Press, 2015).

145   **"Dying Earth":** The genre really picks up steam with H. G. Wells's *The Time Machine*, eventually finding a natural home in postapocalyptic cinema, e.g., *The World, the Flesh, and the Devil* and *The Day After.*

145   **"climate existentialism":** "Nihilism and defeatism in response to the climate crisis isn't either brave or insightful and it's deeply weird to see it treated as some beautiful, poetic intervention," Kate Aronoff has written, on Twitter, referring probably to the writing of Roy Scranton. "Climate change is many things. One thing it's not is a vehicle for literary men to opine on their existential dread and then dress it up as science." See https://twitter.com/KateAronoff/status/1035022145565470725.

146   **literary theorists call metanarrative:** See, especially, Jean-Francois Lyotard, *The Postmodern Condition: A Report on Knowledge* (Minneapolis: University of Minnesota Press, 1984).

146   **as surely as screwball comedies:** A great account of this is Morris Dickstein, *Dancing in the Dark: A Cultural History of the Great Depression* (New York: W. W. Norton, 2009).

146   ***The Great Derangement:*** Ghosh's book (Chicago: University of Chicago Press, 2016) was published with the vivid subtitle *Climate Change and the Unthinkable.*

146   **"cli-fi":** The term has gained currency only over the last decade or so, but examples of the genre—typically speculative fiction driven by climate conditions—date back at least as far as J. G. Ballard (*The Wind from Nowhere, The Drowned World, The Burning World*) and possibly to H. G. Wells (*The Time Machine*) and Jules Verne (*The Purchase of the North Pole*). In other words, it's more or less as old as the science fiction genre, from which it draws its name. Margaret Atwood's *MaddAddam* trilogy (which also includes *The Year of the Flood* and *Oryx and Crake*) surely qualifies, as does even Ian McEwan's *Solar.* All of these test Ghosh's thesis, since they are climate-powered novels with the narrative architecture of the classic bourgeois novel, more or less. Cormac McCarthy's *The Road* is a bit of a different beast—a climate epic. But those who these days talk up cli-fi as a genre seem to mean something more . . . well, *genre*—for instance, Kim Stanley Robinson's *Science in the Capital* trilogy and, later, *New York 2140.* Going back further, J. G. Ballard's *Drowned World* trilogy is an exquisite example.

147   **especially in conventional novels:** Ghosh is dealing here with a very narrow definition of the archetypal novel, emphasizing stories of protagonist
journeys through emerging bourgeois systems. And while he raises the Cold
War and 9/11 as examples of real-world stories that have inspired novels in
that tradition, it's not really the case that the best novels, and films, about the
end of the Cold War are the ones that place their characters very precisely
on a map of the 1989 world, like butterflies pinned to a screen. And the
ones that have approached 9/11 have been mostly duds, as well, though an
entire generation, especially the male half, sometimes seemed to feel called
to literary action by it. "If September 11 had to happen," Martin Amis wrote
in *The Second Plane,* his meditation on the fate of the imagination in the age
of terror, "then I am not at all sorry that it happened in my lifetime." Global
warming has not made Martin Amis feel like George Orwell, as far as I
know, though it has spawned a whole small genre of mourning essay: the
fatalistic, quasi-poetic, first-person ecological lamentation—exemplified by
Roy Scranton, with his *Learning to Die in the Anthropocene* and *We're Doomed.
Now What?*—which may be the closest that climate change stories can get to
the self-mythologizing moral clarity of Orwell.

147   **"man against nature":** This is one of the archetypal "conflict narratives."
Other examples range from *Robinson Crusoe* to *Life of Pi.*

148   **the richest 10 percent:** Oxfam, "Extreme Carbon Inequality," December 2015,
www.oxfam.org/sites/www.oxfam.org/files/file_attachments/mb-extreme
-carbon-inequality-021215-en.pdf.

148   **many on the Left:** The argument is a pervasive one, in part because it is
so persuasive, but has been made with special flair by Naomi Klein in *This
Changes Everything* and *The Battle for Paradise*; Jedediah Purdy in *After Nature*
but perhaps more strikingly in his essays and exchanges published in *Dissent*;
and of course Andreas Malm in *Fossil Capital.*

149   **the socialist countries:** History is not a much better guide, with Left industrialization during Stalin's Five Year Plan or Mao's Great Leap Forward, or
even Venezuela under Hugo Chávez not offering a more responsible approach
than anything that was happening in the West.

149   **The natural villains:** Accounts of the bad behavior of oil companies abound,
too, but two good places to start are Naomi Oreskes and Erik M. Conway,
Merchants of Doubt (New York: Bloomsbury, 2010) and Michael E. Mann
and Tom Toles, *The Madhouse Effect* (New York: Columbia University Press,
2016).

149   **a recent survey of movies:** Peter Kareiva and Valerie Carranza, "Existential
Risk Due to Ecosystem Collapse: Nature Strikes Back," *Futures,* September
2018.

149   **less than 40 percent:** According to the IPCC, the figure is 35 percent: see
IPCC, *Contribution of Working Group III to the Fifth Assessment Report of the
Intergovernmental Panel on Climate Change* (Geneva, 2014).

149   **world's ten biggest oil companies:** Claire Poole, "The World's Largest Oil
and Gas Companies 2018: Royal Dutch Shell Surpasses Exxon as Top Dog,"
Forbes, June 6, 2018.

149   **15 percent of the world's emissions:** According to the World Resources In-

stitute, the figure was 14.36 percent in 2017: Johannes Friedrich, Mengpin Ge, and Andrew Pickens, "This Interactive Chart Explains World's Top Ten Emitters, and How They've Changed," World Resources Institute, April 11, 2017, www.wri.org/blog/2017/04/interactive-chart-explains-worlds-top-10 -emitters-and-how-theyve-changed.

150   **the story of nature and our relationship to it:** In 1980, the art critic John Berger called modern zoos "an epitaph to a relationship that is as old as man": "the zoo to which people go to meet animals, to observe them, is, in fact, a monument to the impossibility of such encounters."

"Today those words could be applied to much of middle-class mass culture," the legal scholar and environmentalist Jedediah Purdy wrote in "Thinking Like a Mountain" (*n+1* 29, Fall 2017), an essay on new forms of nature writing in the age of the Anthropocene. "It has become a kind of memorial to the nonhuman world, revived in a thousand representations even as it disappears all at once." What he means is that we built a zoo out of nature, yes; but we live still inside those cages. "Alongside global domestication, an opposite and terrifying potential broods," Purdy writes. "Every new superstorm, contagion, or annual heat record is pregnant with doom, most acutely for the world's poor, but finally for nearly everyone. For all our deep and accelerating inequalities, life is less dangerous, and the natural world a more stable and fungible backdrop for human activity, than ever before. Yet the whole world also seems poised to come for us like a phalanx of piqued gods who have just switched sides."

150   **half of them extinct:** E. O. Wilson made this prediction in a *New York Times* op-ed, "The Eight Million Species We Don't Know," published on March 3, 2018—and it echoes, conceptually, his 2016 book *Half-Earth: Our Planet's Fight for Life* (New York: W. W. Norton, 2016). According to the 2018 Living Planet report, prepared by the World Wildlife Fund and the Zoological Society of London, world wildlife has already declined that much—in fact, by 60 percent, all since 1970.

151   **Another such parable is bee death:** I wrote a long magazine story about the phenomenon called "The Anxiety of Bees" (*New York,* June 17, 2015).

152   **Flying insects might be disappearing:** The 2017 study was published in *PLOS One* under the unwieldy title "More than 75 Percent Decline over 27 Years in Total Flying Insect Biomass in Protected Areas." In 2018, a survey of insect populations in the rain forests of Puerto Rico was even more alarming—in fact, another researcher called their findings "hyperalarming." Insects there have declined sixtyfold. (Bradford Lister and Andres Garcia, "Climate-Driven Declines in Arthropod Abundance Restructure a Rainforest Food Web," *Proceedings of the National Academy of Sciences,* October 30, 2018.)

152   **devoting whole feature articles:** Jamie Lowe's "The Super Bowl of Beekeeping" (*The New York Times Magazine,* August 15, 2018) is perhaps the most recent example. The original "fable of the bees" had a very different meaning: Bernard Mandeville's 1705 poem of that name was an extended argument that public displays of virtue were invariably hypocritical and that the world was made a better place, in fact, the more ruthlessly individuals pursued their own "vices." That the poem eventually became a touchstone of free-market

thinking, and a major influence on Adam Smith, is all the more remarkable given that it first gained popularity in the aftermath of the South Sea Bubble.

153   **"designer climates":** "If geoengineering worked, whose hand would be on the thermostat?" asked Alan Robock in *Science,* in 2008. "How could the world agree on an optimal climate?" Ten years later, his student Ben Kravitz wrote, on the Harvard geoengineering program's blog—yes, Harvard has a geoengineering program, and yes, they have a blog—"it may be possible to meet multiple, simultaneous objectives in the climate system."

154   **Twenty-two percent:** Jakub Nowosad et al., "Global Assessment and Mapping of Changes in Mesoscale Landscapes: 1992–2015," *International Journal of Applied Earth Observation and Geoinformation* (October 2018).

154   **Ninety-six percent:** Yinon M. Bar-On et al., "The Biomass Distribution on Earth," *Proceedings of the National Academy of the Sciences* (June 2018).

154   **the age of loneliness:** Brooke Jarvis, "The Insect Apocalypse Is Here," *The New York Times Magazine,* November 27, 2018.

155   **"scientific reticence":** J. E. Hansen, "Scientific Reticence and Sea Level Rise," *Environmental Research Letters* 2 (May 2007).

157   **a 2017 *Nature* paper:** Daniel A. Chapman et al., "Reassessing Emotion in Climate Change Communication," *Nature Climate Change* (November 2017): pp. 850–52.

157   **IPCC released a dramatic, alarmist report:** IPCC, *Global Warming of 1.5°C: An IPCC Special Report on the Impacts of Global Warming of 1.5°C Above Pre-Industrial Levels and Related Global Greenhouse Gas Emission Pathways, in the Context of Strengthening the Global Response to the Threat of Climate Change, Sustainable Development, and Efforts to Eradicate Poverty* (Incheon, Korea, 2018), www.ipcc.ch/report/sr15.

Crisis Capitalism

158   **The scroll of cognitive biases:** The single best primer on what behavioral economics has to teach us about these biases is by the Nobel laureate Daniel Kahneman, *Thinking, Fast and Slow* (New York: Farrar, Straus & Giroux, 2013).

160   **the scope of the climate threat:** This is why the theorist Timothy Morton refers to climate change as a "hyperobject." But while the term is useful in suggesting just how large climate change is, and just how poorly we've been able to perceive that scale to date, the deeper you get into Morton's analysis, the less illuminating it becomes. In *Hyperobjects: Philosophy and Ecology After the End of the World* (Minneapolis: University of Minnesota Press, 2013), he names five characteristics: hyperobjects are 1) *viscous,* by which he means that they stick to any object or idea they come into contact with, like oil; 2) *molten,* by which he means so big, they seem to defy our sense of space-time; 3) *nonlocal,* by which he means distributed in ways that frustrate any attempt to perceive them entirely from a single perspective; 4) *phased,* by which he means that they have dimensional qualities we cannot understand, as we wouldn't understand a five-dimensional object passing through our three-dimensional space; and 5) *interobjective,* by which he means that they connect divergent items and systems. Viscous, nonlocal, and interobjective—okay. But these do not make global warming a different kind of

phenomenon than we have seen before, or than those—like capitalism, say—that we actually understand quite well. As for the other qualities . . . If climate change defies our sense of space-time, it is only because we have an impoverished, narrow idea of space-time, since in fact warming is taking place very much within our planet's atmosphere, not inexplicably but in ways scientists have predicted quite precisely over decades. That we have failed to deal with it, over those same decades, does not mean it is literally beyond our comprehension. Saying so sounds almost like a cop-out, in fact.

161   **"It is easier to imagine":** Jameson wrote this in "Future City," published in *New Left Review* in May–June 2003.

162   **pet theory of the socialist Left:** Degrees of emphasis vary, of course, but you can find forms of the "fossil capitalism" argument in Vaclav Smil's *Energy and Civilization,* along with Andreas Malm's *Fossil Capital* and Jason Moore's *Capitalism in the Web of Life.*

162   **Can capitalism survive climate change?:** Moore raises this question in *Capitalism in the Web of Life,* and it is discussed at some length in Benjamin Kunkel, "The Capitalocene," *London Review of Books,* March 2, 2017.

163   **Klein memorably sketched out:** Naomi Klein, *The Shock Doctrine: The Rise of Disaster Capitalism* (New York: Picador, 2007).

163   **the island of Puerto Rico:** Naomi Klein, *The Battle for Paradise: Puerto Rico Takes On the Disaster Capitalists* (Chicago: Haymarket, 2018).

163   **Maria could cut Puerto Rican incomes:** This comes from Hsiang and Houser's "Don't Let Puerto Rico Fall into an Economic Abyss," *The New York Times,* September 29, 2017.

165   **carbon emissions have exploded:** According to the International Energy Agency, total global emissions were 32.5 gigatons in 2017, up from 22.4 in 1990. Of course, it is worth remembering that the world's socialist nations, and even its left-of-center ones, do not have a notably better record when it comes to emissions than its excessively capitalistic ones. This suggests that it may be a bit misleading to describe emissions as having been driven by capitalism per se, or even interests made especially prominent and powerful within capitalistic systems. Instead, it may reflect the universal power of material comforts, benefits that we tend to assess using only a very short-term calculus.

165   **"Neoliberalism: Oversold?":** This paper, by Jonathan D. Ostry, Prakash Loungani, and Davide Furceri, was published in June 2016.

165   **something like a fantasy field:** Romer published "The Trouble with Macroeconomics" on his own website on September 14, 2016.

166   **Nordhaus favors a carbon tax:** The Nobel laureate has published widely on the subject of the carbon tax, and he gives the most plainspoken account of the tax level he considers optimal in "Integrated Assessment Models of Climate Change," National Bureau of Economic Research, 2017, https://www.nber.org/reporter/2017number3/nordhaus.html.

166   **$306 billion:** Adam B. Smith, "2017 U.S. Billion-Dollar Weather and Climate Disasters: A Historic Year in Context," National Oceanic and Atmospheric Association, January 8, 2018.

166   **$551 trillion in damages:** "Risks Associated with Global Warming of 1.5 De-

grees Celsius or 2 Degrees Celsius," Tyndall Centre for Climate Change Research, May 2018.

166   **23 percent of potential global income:** Marshall Burke et al., "Global Non-Linear Effect of Temperature on Economic Production," *Nature* 527 (October 2015): pp. 235–39, https://doi.org/10.1038/nature15725.

169   **Of 400 IPCC emissions models:** "Negative Emissions Technologies: What Role in Meeting Paris Agreement Targets?" European Academies' Science Advisory Council, February 2018.

169   **a third of the world's farmable land:** Jason Hickel, "The Paris Agreement Is Deeply Flawed—It's Time for a New Deal," *Al Jazeera*, March 16, 2018.

170   **a paper by David Keith:** David Keith et al., "A Process for Capturing CO_2 from the Atmosphere," *Joule*, August 15, 2018.

170   **total global fossil fuel subsidies:** David Coady et al., "How Large Are Global Fossil Fuel Subsidies?" *World Development* 91 (March 2017): pp. 11–27.

170   **$2.3 trillion tax cut:** David Rogers, "At $2.3 Trillion Cost, Trump Tax Cuts Leave Big Gap," *Politico*, February 28, 2018. Other estimates run higher.

The Church of Technology

171   **outlined by Eric Schmidt:** He laid out this perspective most clearly at a conference in New York in January 2016.

172   **"Consider: Who pursues":** Ted Chiang, "Silicon Valley Is Turning into Its Own Worst Fear," BuzzFeed, December 18, 2017.

173   **an influential 2002 paper:** Nick Bostrom, "Analyzing Human Extinction Scenarios and Related Hazards," *Journal of Evolution and Technology* 9 (March 2002).

174   **close to universal:** In "Survival of the Richest" (*Medium,* July 5, 2018), the futurist Douglas Rushkoff described his experience as a keynote speaker at a private conference attended by the superrich—these patrons not themselves technologists but hedge-funders he came to feel were taking all of their cues from them. Quickly, he writes, the conversation attained a clear focus:

> Which region will be less impacted by the coming climate crisis: New Zealand or Alaska? Is Google really building Ray Kurzweil a home for his brain, and will his consciousness live through the transition, or will it die and be reborn as a whole new one? Finally, the CEO of a brokerage house explained that he had nearly completed building his own underground bunker system and asked, "How do I maintain authority over my security force after the event?"

"The event." In Rushkoff's telling, this is a kind of catchall phrase for anything that might threaten their status or security as the world's most privileged—"their euphemism for the environmental collapse, social unrest, nuclear explosion, unstoppable virus, or Mr. Robot hack that takes everything down.

"This single question occupied us for the rest of the hour," Rushkoff continues.

> They knew armed guards would be required to protect their compounds from the angry mobs. But how would they pay the guards once money was worthless? What would stop the guards from choosing their own leader? The billionaires considered using special combination locks on the food supply that only they knew. Or making guards wear disciplinary collars of some kind in return for their survival. Or maybe building robots to serve as guards and workers—if that technology could be developed in time.

In *To Be a Machine,* Mark O'Connell traced the same impulse through Silicon Valley's whole Brahman caste. The book opens with an epigraph from Don DeLillo: "This is the whole point of technology. It creates an appetite for immortality on the one hand. It threatens universal extinction on the other." The quote comes from *White Noise,* in particular from its narrator's colleague and sidekick Murray Jay Siskind, who is both the novel's comic foil and its "explainer." It was never clear to me just how seriously we are meant to take Murray's pronouncements, but this one does quite sharply describe the contemporary tech two-step: freaking out about "existential risks" while simultaneously cultivating private exits from mortality.

For Rushkoff, these are all facets of the same impulse, broadly shared by the class of visionaries and power brokers and venture capitalists whose dreams for the future are received as blueprints, especially by the armies of engineers they command like impetuous fiefdoms—investing in new forms of space travel, life extension, and technology-aided life after death. "They were preparing for a digital future that had a whole lot less to do with making the world a better place than it did with transcending the human condition altogether and insulating themselves from a very real and present danger of climate change, rising sea levels, mass migrations, global pandemics, nativist panic, and resource depletion," he writes. "For them, the future of technology is really about just one thing: escape."

176   **"An Account of My Hut"**: Christina Nichol, "An Account of My Hut," *n+1,* Spring 2018. Nichol explains the title this way:

> I once read a story called "An Account of My Hut," by Kamo no Chōmei, a 12th-century Japanese hermit. Chōmei describes how after witnessing a fire, an earthquake, and a typhoon in Kyoto, he leaves society and goes to live in a hut.

> Seven hundred years later, Basil Bunting, the Northumberland poet, wrote his own rendition of Chōmei's story:

> > *Oh! There's nothing to complain about.*
> > *Buddha says: "None of the world is good."*
> > *I am fond of my hut . . .*

> But even if I wanted to renounce the world, I wouldn't be able to afford a hut in California.

177   **as old as John Maynard Keynes:** Keynes extended the prediction—much, much talked about ever since—in an essay notably published in 1930, just after the stock market crash of 1929: John Maynard Keynes, "Economic Possibilities for Our Grandchildren," *Nation and Athenaeum*, October 11 and 18, 1930.

177   **"You can see the computer age":** This line first appeared in Robert M. Solow, "We'd Better Watch Out," review of *Manufacturing Matters* by Stephen S. Cohen and John Zysman, *The New York Times Book Review*, July 12, 1987.

179   **a million transatlantic flights:** Alex Hern, "Bitcoin's Energy Usage Is Huge—We Can't Afford to Ignore It," *The Guardian*, January 17, 2018.

179   **"If we don't act quickly":** Bill McKibben, "Winning Is the Same as Losing," *Rolling Stone,* December 1, 2017. "Another way of saying this: By 2075, the world will be powered by solar panels and windmills—free energy is a hard business proposition to beat," McKibben wrote. "But on current trajectories, they'll light up a busted planet. The decisions we make in 2075 won't matter; indeed, the decisions we make in 2025 will matter much less than the ones we make in the next few years. The leverage is now."

179   **"The future is already here":** The quip first appeared in *The Economist* in 2003.

179   **less than 10 percent of the world:** IDC, "Smartphone OS Market Share," www.idc.com/promo/smartphone-market-share/os.

179   **somewhere between a quarter and a third:** David Murphy, "2.4BN Smartphone Users in 2017, Says eMarketer," *Mobile Marketing*, April 28, 2017, https://mobilemarketingmagazine.com/24bn-smartphone-users-in-2017-says-emarketer.

179   **global decarbonization in 2000:** These figures come from Robbie Andrew, a senior researcher at the Center for International Climate Research, and his presentation "Global Collective Effort," which he published on his website in May 2018 (http://folk.uio.no/roberan/t/2C.shtml). He was drawing on figures put forward by Michael R. Raupach et al. in "Sharing a Quota on Cumulative Carbon Emissions," *Nature Climate Change* (September 2014).

180   **only one year:** "UN Secretary-General Antonio Guterres Calls for Climate Leadership, Outlines Expectations for Next Three Years," *UN Climate Change News*, September 10, 2018: "If we do not change course by 2020, we risk missing the point where we can avoid runaway climate change, with disastrous consequences for people and all the natural systems that sustain us."

180   **poured more concrete in three years:** Jocelyn Timperley, "Q&A: Why Cement Emissions Matter for Climate Change," *Carbon Brief,* September 13, 2018, www.carbonbrief.org/qa-why-cement-emissions-matter-for-climate-change.

181   **the world would need to add:** Ken Caldeira, "Climate Sensitivity Uncertainty and the Need for Energy Without CO_2 Emission," *Science* 299 (March 2003): pp. 2052–54.

181   **in four hundred years:** James Temple, "At This Rate, It's Going to Take Nearly 400 Years to Transform the Energy System," *MIT Technology Review*, March 14, 2018, www.technologyreview.com/s/610457/at-this-rate-its-going-to-take-nearly-400-years-to-transform-the-energy-system.

183   **official death count is 47:** U.N. Information Service, "New Report on Health

Effects Due to Radiation from the Chernobyl Accident," February 28, 2011, www.unis.unvienna.org/unis/en/pressrels/2011/unisinf398.html.

183   **as high as 4,000:** World Health Organization, "Chernobyl: The True Scale of the Accident," September 5, 2005, www.who.int/mediacentre/news/releases/2005/pr38.

183   **"no discernible increased incidence":** United Nations, "Report of the United Nations Scientific Committee on the Effects of Atomic Radiation" (May 2013): p. 11, www.unscear.org/docs/GAreports/A-68-46_e_V1385727.pdf.

183   **an additional 1,400 Americans:** Lisa Friedman, "Cost of New E.P.A. Coal Rules: Up to 1,400 More Deaths a Year," *The New York Times*, August 21, 2018.

183   **nine million each year:** Pamela Das and Richard Horton, "Pollution, Health, and the Planet: Time for Decisive Action," *The Lancet* 391, no. 10119 (October 2017): pp. 407–8, https://doi.org/10.1016/S0140-6736(17)32588-6.

184   **growing its carbon emissions:** James Conca, "Why Aren't Renewables Decreasing Germany's Carbon Emissions?" *Forbes*, October 10, 2017.

184   **"How many will play augmented reality games":** Andreas Malm, *The Progress of This Storm: Nature and Society in a Warming World* (London: Verso, 2018).

184   **The poet and musician Kate Tempest:** These are lyrics to her song "Tunnel Vision."

Politics of Consumption

185   **a note, handwritten:** Annie Correal, "What Drove a Man to Set Himself on Fire in Brooklyn?" *The New York Times*, May 28, 2018.

185   **a longer letter, typed:** For an in-depth account of this letter, see Theodore Parisienne et al., "Famed Gay Rights Lawyer Sets Himself on Fire at Prospect Park in Protest Suicide Against Fossil Fuels," New York *Daily News*, April 14, 2018.

187   **moral arms race:** Citizens who now clean their consciences with philanthropic donations directed toward medical research, college scholarships, or museums and literary magazines may begin increasingly to do so by buying carbon offsets or investing in carbon-capture funds (indeed, some progressive nations may invest the proceeds of carbon taxes directly into CCS and BECCS). Progressive scientists will apply gene therapy to climate change, as they have already begun to do with the woolly mammoth—which they hope, once brought back to life, might restore the grasslands of the Eurasian steppe and prevent methane release from permafrost—and will probably do soon with the mosquito, hoping to eradicate mosquito-borne disease. Perhaps a rogue billionaire will try to single-handedly cool the earth with geoengineering, flying a few private planes around the equator to disperse sulfur and citing the model of Bill Gates and his mosquito nets.

187   **"apparatus of justification":** Thomas Piketty, *Capital in the Twenty-First Century* (Cambridge, MA: Harvard University Press, 2014).

188   **SoulCycle, Goop, Moon Juice:** The founder of hipster foodie magazine *Modern Farmer* is, in 2018, rumored to be launching a "Goop for climate change."

188   **the pesticide Roundup:** Alexis Temkin, "Breakfast with a Dose of Roundup?" Environmental Working Group Children's Health Initiative, August 15, 2018, www.ewg.org/childrenshealth/glyphosateincereal.

188   **elaborate guidance:** "During a wildfire, dust masks aren't enough!" the National Weather Service warned on Facebook. "They won't protect you from the fine particles in wildfire smoke. It is best to stay indoors, keeping windows and doors closed. If you're running an air conditioner, keep the fresh-air intake closed and the filter clean to prevent outdoor smoke from getting inside."

189   **"philanthrocapitalism":** Perhaps the most piercing account of this phenomenon is Anand Giridharadas, *Winners Take All: The Elite Charade of Changing the World* (New York: Knopf, 2018).

190   **"moral economy":** This story is recounted in Tim Rogan, *The Moral Economists* (Princeton, NJ: Princeton University Press, 2018); see also Tehila Sasson's review, published in *Dissent* under the headline "The Gospel of Wealth," August 22, 2018.

190   **asked to be entrepreneurs:** Stephen Metcalf, among many others, has written memorably about this phenomenon, in his brief history of neoliberalism, "Neoliberalism: The Idea That Swallowed the World," *The Guardian*, August 18, 2017.

191   *Climate Leviathan:* Geoff Mann and Joel Wainwright, *Climate Leviathan: A Political Theory of Our Planetary Future* (London: Verso, 2018).

194   **In 2018, an illuminating study:** Katharine Ricke et al., "Country-Level Social Cost of Carbon," *Nature Climate Change* 8 (September 2018): pp. 895–900.

195   **Belt and Road Initiative:** Perhaps the best account of this initiative is Bruno Maçães's *Belt and Road: A Chinese World Order* (London: Hurst, 2018). The initiative "may also promote permanent environmental degradation," a group of researchers argued recently. (Fernando Ascensão et al., "Environmental Challenges for the Belt and Road Initiative," *Nature Sustainability*, May 2018).

195   **the possibility of disequilibrium:** Harald Welzer, *Climate Wars: What People Will Be Killed For in the 21st Century* (Cambridge: Polity, 2012).

196   **pulls criminals out of pop concerts:** According to *The Washington Post*'s Hamza Shaban, this happened three times in just two months in the spring of 2018: "Facial Recognition Cameras in China Snag Man Who Allegedly Stole $17,000 Worth of Potatoes," May 22, 2018.

196   **domestic-spy drones:** Stephen Chen, "China Takes Surveillance to New Heights with Flock of Robotic Doves, but Do They Come in Peace?" *South China Morning Post*, June 24, 2018.

History After Progress

197   **most unshakable creeds:** It wasn't just the promise of growth that was invented in the industrial era, but the idea of history, which promises that the past tells a story of human progress—and suggests, therefore, that the future will, too.

This progressive faith has a demotic basis, which is that daily life changed so quickly in the Victorian era that no one with eyes open could have missed it. It also has an intellectual one, which is that philosophers from Hegel to Comte proposed, at various points in the nineteenth century, that history had a shape—that it evolved, in one form or another, toward the light, of one kind or another. The idea would not have confused readers of their contemporaries Darwin and Spencer. Nor, for that matter, visitors to Queen Victoria's Crystal Palace exhibition, the first World's Fair, which organized national showcases into an implicit competition of relative development and more or less promised that technology would bring about a better future for all. By the time Jacob Burckhardt was writing his *Civilization of the Renaissance in Italy*, which furnished the now-proverbial three-act structure of Western history—antiquity followed by the Dark Ages followed by modernity—he could imagine himself as an opponent of both Hegel and Comte and yet nevertheless produce a work that explicitly periodized the past into a single unfolding drama. That is how thoroughly the idea of progressive history had taken hold in a time of rapid social, economic, and cultural change: even critics of reflexive Western triumphalism tended to see history as marching forward. Marx is the clearest example: squint at his reimagined Hegelianism, and its shape looks a lot like the enduring wall-chart of history first published by Sebastian Adams—motivated by Christian evangelism, amazingly—in 1871. In 1920, H. G. Wells published his influential version, *The Outline of History*; in it, he declared that "the history of mankind," which he traced through forty chapters from "The Earth in Space and Time" to "The Next Stage of History," "is a history of more or less blind endeavours to conceive a common purpose in relation to which all men may live happily." It sold millions of copies and was translated into dozens of languages, and it casts a shadow over nearly every project of popular, long view history undertaken since, from Kenneth Clark's *Civilisation* to Jared Diamond's *Guns, Germs, and Steel*.

198   *Sapiens*: That this kind of total skepticism won Harari such an admiring audience among so many leading avatars of technocratic progress is one of the curiosities of the TED Talk age. But the skepticism also flatters, especially those inclined by their own sense of accomplishment to contemplate the longest sweeps of history. Inviting you to contemplate that history, Harari also seems to pull you beyond or outside it. In this way, he shares strains of lecturesome DNA not just with Diamond but with Joseph Campbell and even Jordan Peterson. In his subsequent book, *Homo Deus*, Harari endorses a new contemporary myth, though he doesn't quite recognize it as such—making his own case for the near-term arrival of a superpowerful artificial intelligence that will render everything we know as "humanity" close to obsolete.

198   *Against the Grain*: The human remains excavated from this time tell a clear story of human strife: the people were shorter, sicker, and died younger than their predecessors. The average height fell from 5'10" for men and 5'6" for women to 5'5" and 5'1", respectively; settled communities were more vulnerable to infectious disease, but obesity and heart disease also spiked. This is why it's so that the case against civilization, as the critic John Lanchester has called it, can be made simply as a case against farming.

198   **"the worst mistake"**: Jared Diamond, "The Worst Mistake in the History of the Human Race," *Discover*, May 1987.

200   **"we might call the Liberal Story"**: Yuval Noah Harari, "Does Trump's Rise Mean Liberalism's End?" *The New Yorker*, October 7, 2016.

201   **ekpyrosis**: This was the belief that, periodically, the cosmos would be entirely destroyed in what was called a "Great Year," then be re-created and the process would begin again. Plato preferred the term "perfect year," in which the stars would be returned to their original positions.

201   **"dynastic cycle"**: Although some accounts of the cycle offered a dozen or more phases, according to the Chinese philosopher Mencius the cycle had only three (essentially rise, peak, and decline).

201   **"eternal recurrence"**: Nietzsche first proposes this idea, that everything is bound to repeat itself eternally, as a sort of thought experiment in *The Gay Science* (1882). But he would return to it again and again, often describing it as something more like a law of the universe—which is similar to how it was treated by the ancient Egyptians, Indians, and Greek Stoics.

201   **"public purpose" and "private interest"**: Arthur M. Schlesinger, *The Cycles of American History* (New York: Houghton Mifflin, 1986).

201   **The Rise and Fall of the Great Powers**: In his 1987 book, Kennedy offers a relatively simple model of great-power history: growth fueled by natural resources followed by decline precipitated by military overreach.

202   **The Progress of This Storm**: The main thrust of this book, Malm's follow-up to *Fossil Capital*, is that while we may believe that "nature," as something distinct from "society," has disappeared, in fact global warming has brought it back with a punitive vengeance.

Ethics at the End of the World

207   **podcast "S-Town"**: McLemore, whose panic may have been caused in part by mercury poisoning, was most concerned about Arctic ice melt, drought, and the slowdown of the thermohaline convector.

207   **"I sometimes call it toxic knowledge"**: Richard Heinberg, "Surviving S-Town," Post Carbon Institute, April 7, 2017.

207   **"nature is thriving"**: Thomas's book is *Inheritors of the Earth: How Nature Is Thriving in an Age of Extinction* (New York: Public Affairs, 2017), and while it offers not so much a full-throated celebration of what he calls an "age of extinction" but a more modest proposal that we view the positive, generative effects of climate change alongside its crueler impacts. This is a note of contrarian optimism echoing Michael Shellenberger and Ted Nordhaus, in their *Break Through: Why We Can't Leave Saving the Planet to Environmentalists* and *Love Your Monsters: Postenvironmentalism and the Anthropocene*; and the Canadian, Swedish, and South African academics behind the research collaboration "Bright Spots," who, despite considerably more concern about the effects of global warming, nevertheless keep a running list of positive environmental developments they believe makes the case for what they call a "good Anthropocene."

208   **"The Second Coming"**: Among other things, Yeats gave Joan Didion the

lines she built into her essay "Slouching Towards Bethlehem": "Things fall apart; the centre cannot hold; / Mere anarchy is loosed upon the world."

209   **"immanent anti-humanism"**: The program is also neatly contained in Jeffers's most famous poem, "Carmel Point":

> *We must uncenter our minds from ourselves;*
> *We must unhumanize our views a little, and become confident*
> *As the rock and ocean that we were made from.*

210   **a time of approaching ecological collapse**: Indeed, the manifesto continues, "human civilization is an intensely fragile construction," and yet, they write, we are forever in denial about that fragility—our very day-to-day lives depend on that denial of fragility, perhaps as much as they depend on the denial of our own mortality. This is what the philosopher Samuel Scheffler means when he suggests that, in an agnostic world, the role once played by an afterlife in inspiring and organizing and policing moral and ethical behavior has been taken up, in part, by the conviction that the world will continue on after us when we die. In other words, the idea that life is not just worth living but worth living well, he suggests, "would be more threatened by the prospect of humanity's disappearance than by the prospect of our own deaths." As Charles Mann summarizes Scheffler, considering the ethical paradox of human action on climate change, "The belief that human life will continue, even if we ourselves die, is one of the underpinnings of society."

"Once that belief begins to crumble, the collapse of a civilization may become unstoppable," Kingsnorth and Hine wrote in their manifesto. "That civilizations fall, sooner or later, is as much a law of history as gravity is a law of physics. What remains after the fall is a wild mixture of cultural debris, confused and angry people whose certainties have betrayed them, and those forces which were always there, deeper than the foundations of the city walls: the desire to survive and the desire for meaning."

210   **"We believe that the roots"**: "We do not believe that everything will be fine," Kingsnorth and Hine write. "We are not even sure, based on current definitions of progress and improvement, that we want it to be."

In the manifesto, Dark Mountain outlined what they called "the eight principles of uncivilization," a sort of mission statement for their movement that moves from general principle and perceptions to a more focused statement of intent. "We reject the faith which holds that the converging crises of our time can be reduced to a set of 'problems' in need of technological or political 'solutions,'" the list begins, and though they foreswear these kinds of solutions, they don't entirely give up on response. But Dark Mountain is ultimately a literary collective—organizing festivals, workshops, and meditation retreats—and the most concrete, practical response they call for in their manifesto is in art. "We believe that the roots of these crises lie in the stories we have been telling ourselves," namely "the myth of progress, the myth of human centrality, and the myth of our separation from 'nature.'" These, they add, "are more dangerous for the fact that we have forgotten they are myths." In response, they promise, "we will assert the role of storytelling as more than mere entertainment" and "will write with dirt under our fingernails."

The goal: through storytelling, to find a new vantage from which the end of civilization would not seem so bad. In a certain way, they suggest, they themselves have already achieved this state of enlightenment. "The end of the world as we know it is not the end of the world, full stop," they write. "Together we will find the hope beyond hope, the paths which lead to the unknown world ahead of us."

210   **Kingsnorth published a new manifesto:** Paul Kingsnorth, "Dark Ecology," *Orion*, November–December 2012. This manifesto includes this passage:

> What does the near future look like? I'd put my bets on a strange and unworldly combination of ongoing collapse, which will continue to fragment both nature and culture, and a new wave of techno-green "solutions" being unveiled in a doomed attempt to prevent it. I don't believe now that anything can break this cycle, barring some kind of reset: the kind that we have seen many times before in human history. Some kind of fall back down to a lower level of civilizational complexity. Something like the storm that is now visibly brewing all around us.
>
> If you don't like any of this, but you know you can't stop it, where does it leave you? The answer is that it leaves you with an obligation to be honest about where you are in history's great cycle, and what you have the power to do and what you don't. If you think you can magic us out of the progress trap with new ideas or new technologies, you are wasting your time. If you think that the usual "campaigning" behavior is going to work today where it didn't work yesterday, you will be wasting your time. If you think the machine can be reformed, tamed, or defanged, you will be wasting your time. If you draw up a great big plan for a better world based on science and rational argument, you will be wasting your time. If you try to live in the past, you will be wasting your time. If you romanticize hunting and gathering or send bombs to computer store owners, you will be wasting your time.

213   **tending toward more engagement:** You can see this in how quite radical thinkers about the environment and our obligations to it, from Jedediah Purdy to Naomi Klein, focus so intently on the problems of political action. In Purdy's *After Nature: A Politics for the Anthropocene* (Cambridge, MA.: Harvard University Press, 2015), he builds an entire practical politics out of the intuition, inarguably true, that the final and total conquest of the planet by people is marked simultaneously by its degradation; and argues that the end of that long era of natural abundance demands a more democratic approach to environmental politics, policy, and law—even when, or perhaps especially because, any alteration from the present course seems almost infrastructurally impossible. In a 2017 exchange with Katrina Forrester, later published in *Dissent,* he elaborated:

> Here is our paradox: The world can't go on this way; *and* it can't do otherwise. It was the collective power of some—not all—human beings that got us into this: power over resources, power over the

seasons, power over one another. That power has created a global humanity, entangled in a Frankenstein ecology. But it does not yet include the power of accountability or restraint, the power we need. To face the Anthropocene, humans would need a way of facing one another. We would need, first, to *be* a *we*.

From a certain vantage, this may just look like conventional politics, of the kind that Kingsnorth derides as impossibly naive. They are also my politics, for what it's worth—I nod my head in recognition when I read Kate Marvel calling for courage rather than hope, and when I read Naomi Klein rhapsodizing about a community of political resistance growing out of the local sites of protests she calls "Blockadia." I believe, as Purdy does, that the degradation of the planet and the end of natural abundance demand a new progressivism animated by a renewed egalitarian energy; and I believe, as Al Gore does, that we should push technology to chase every last glimmer of hope for averting disastrous climate change—including unleashing, or indulging, market forces to help do so when we can. I believe, as Klein does, that some particular market forces have almost conquered our politics, but not entirely, leaving a bright shining sliver of opportunity; and I also believe, as Bill McKibben does, that meaningful and even dramatic change can be achieved through the familiar paths: voting and organizing and political activity deployed at every level. In other words, I believe in *engagement* above all, engagement wherever it may help. In fact, I find any other response to the climate crisis morally incomprehensible.

213   **global mobilization:** That this is a familiar analogy is unfortunate, because it blunts the intended impression: the Allies' mobilization was unprecedented in human history and has never been matched since. We did not defeat the Nazis with a change to the marginal tax rate, as much as proponents of a climate tax want to see it as a single-step cure-all. In World War II, there was also a draft, a nationalization of industry, and widespread rationing. If you can imagine a tax rate on carbon that would produce that kind of action in just three decades, you have a better imagination than I do.

214   **"eco-nihilism":** Wendy Lynne Lee, *Eco-Nihilism: The Philosophical Geopolitics of the Climate Change Apocalypse* (Lanham, MD: Lexington, 2017).

214   **"climate nihilism":** Parker used the term in explaining his decision to quit Canada's New Democratic Party after its premier endorsed subsidizing natural gas.

214   **"climatic regime":** In an essay titled "Love Your Monsters," Latour elaborated a jeremiad of environmental responsibility from Mary Shelley's parable, one that begins with a perhaps romantic plea for a clear-eyed recognition of just what we have wrought—writing that, "just as we have forgotten that Frankenstein was the man, not the monster, we have also forgotten Frankenstein's real sin."

> Dr. Frankenstein's crime was not that he invented a creature through some combination of hubris and high technology, but rather that he *abandoned the creature to itself.* When Dr. Frankenstein meets his creation on a glacier in the Alps, the monster

claims that it was not *born* a monster, but that it became a criminal only *after* being left alone by his horrified creator, who fled the laboratory once the horrible thing twitched to life.

A similar case for responsibility comes from Donna Haraway, the theorist behind the pioneering feminist *Cyborg Manifesto* (1985), in her more recent *Staying with the Trouble*, subtitled *Making Kin in the Chthulucene* (Durham, NC: Duke University Press, 2016)—after Chthulu, H. P. Lovecraft's many-faced monster of cosmic malevolence.

214   **"human futilitarianism"**: Sam Kriss and Ellie Mae O'Hagan, "Tropical Depressions," *The Baffler* 36 (September 2017). "Climate change means, quite plausibly, the end of everything we now understand to constitute our humanity," Kriss and O'Hagan write. "Something about the magnitude of all this is shattering: most people try not to think about it too much because it's unthinkable, in the same way that death is always unthinkable for the living. For the people who have to think about it—climate scientists, activists, and advocates—that looming catastrophe evokes a similar horror: the potential extinction of humanity in the future puts humanity into question now."

215   **"species loneliness"**: "If the most common causes of individual suicide are depression and psychic isolation, the cause of our accelerating and collectively willed suicide may be despair over the failed system of capitalism and commodity-driven meaning, as well as the crippling condition that psychologists call 'species loneliness,'" Powers told Everett Hamner of *The Los Angeles Review of Books* (April 7, 2018), in an interview published under the headline "Here's to Unsuicide." "We will always be parasites on plants. But that parasitism can be turned into something better—a mutualism. One of my radicalized activists makes this proposal: We should cut trees like they are a gift, not like they are something we a priori deserve. Such a shift in consciousness might have the effect of slowing down deforestation, since we tend to care for gifts better than we do for freebies. But it would also go a long way toward treating the suicidal impulse in people caused by species loneliness. Many indigenous people knew this for millennia: thanking a living thing and asking for its pardon before using it goes a long way toward exonerating the guilt that leads to violence against the self and others."

The Anthropic Principle

219   **rudimentary understanding:** Eunice Foote, "Circumstances Affecting the Heat of the Sun's Rays," *The American Journal of Science and Arts* 22, no. 46 (November 1856). This paper, in which Foote describes the effect of carbon dioxide on global temperature, was first presented at a meeting of the American Association for the Advancement of Science in 1856—where it was read by a male colleague, Joseph Henry. John Tyndall published his work several years later, in 1859.

221   **"Where is everybody?":** In 1985, Los Alamos published a history of the conversation; see Eric M. Jones, "Where Is Everybody?: An Account of Fermi's Question," www.osti.gov/servlets/purl/5746675.

221   **For the entire historical window:** Perhaps the most vivid illustration of this is the xkcd web comic "A Timeline of Earth's Average Temperature," September 12, 2016.

221   **"the great filter":** Hanson first published his thinking on this subject in a 1998 paper, the ominous last line of which is "If we can't find the Great Filter in our past, we'll have to fear it in our future." Robert Hanson, "The Great Filter—Are We Almost Past It?" September 15, 1998, http://mason.gmu.edu/~rhanson/greatfilter.html.

222   **"Is it inhabited?":** This comes from Archibald MacLeish's beautiful account, published on the front page of *The New York Times*, December 25, 1968—the day after *Apollo 8* orbited the moon—under the headline "Riders on Earth Together, Brothers in Eternal Cold." The case MacLeish made was that seeing the planet from a distance could profoundly change how we saw our place in the universe: "Men's conception of themselves and of each other has always depended on their notion of the earth," he wrote.

> Now, in the last few hours, the notion may have changed again. For the first time in all of time men have seen it not as continents or oceans from the little distance of a hundred miles or two or three, but seen it from the depth of space; seen it whole and round and beautiful and small as even Dante—that "first imagination of Christendom"—had never dreamed of seeing it; as the Twentieth Century philosophers of absurdity and despair were incapable of guessing that it might be seen. And seeing it so, one question came to the minds of those who looked at it. "Is it inhabited?" they said to each other and laughed—and then they did not laugh. What came to their minds a hundred thousand miles and more into space—"half way to the moon" they put it—what came to their minds was the life on that little, lonely, floating planet; that tiny raft in the enormous, empty night. "Is it inhabited?"
>
> The medieval notion of the earth put man at the center of everything. The nuclear notion of the earth put him nowhere—beyond the range of reason even—lost in absurdity and war. This latest notion may have other consequences. Formed as it was in the minds of heroic voyagers who were also men, it may remake our image of mankind. No longer that preposterous figure at the center, no longer that degraded and degrading victim off at the margins of reality and blind with blood, man may at last become himself.

222   **the Drake equation:** Drake himself saw the equation as something very preliminary and tentative, a list of factors that would influence the likelihood of finding extraterrestrial intelligence, which he sketched out in advance of a small conference to discuss the issue in 1960. In 2003, Drake recounted the story in *Astrobiology Magazine* under the title "The Drake Equation Revisited" (September 29, 2003).

223   **literally closed themselves off:** Dyson first proposed this possibility in a 1960 paper, "Search for Artificial Stellar Sources of Infrared Radiation" (*Science*

131, no. 3414 [June 1960], pp. 1667–68), though as a concept it appeared earlier in the 1937 sci-fi novel *Star Maker* by Olaf Stapledon.

223   **"the astrobiology of the Anthropocene"**: Adam Frank, *Light of the Stars: Alien Worlds and the Fate of the Earth* (New York: W. W. Norton, 2018). In this book, Frank writes, "Our technology and the vast energies it has unleashed give us enormous power over ourselves and the world around us. It's like we've been given the keys to the planet. Now we're ready to drive it off a cliff."

223   **"thinking like a planet"**: The phrase also recalls Aldo Leopold's "thinking like a mountain," which first appeared in his *Sand County Almanac* of 1937, and which provided the title of an excellent meditative essay by Jedediah Purdy on nature writing and our changing relationship to the natural world, published in *n+1* in 2017.

    Personally the perspective strikes me as too Stoic—a mountain would not much care if humans, a single species, suffered tremendous setbacks, and the same is true for the planet as a whole. As those scientists kept reminding me, "The earth will survive; it's the humans that might not." And indeed, commentators have traced a prehistory of Leopold's phrase through the ancient philosophy of Epicurus and Lucretius.

223   **an unconventional recent paper:** Gavin A. Schmidt, "The Silurian Hypothesis: Would It Be Possible to Detect an Industrial Civilization in the Geological Record?" *International Journal of Astrobiology*, April 16, 2018, https://doi.org/10.1017/S1473550418000095.

224   **anyone trying to "solve" the Drake equation:** One especially notable effort was Anders Sandberg's et al., "Dissolving the Fermi Paradox," Future of Humanity Institute, Oxford University, June 6, 2018, https://arxiv.org/pdf/1806.02404.pdf.

226   **"Now I am become death":** An account of this—including the fact that Oppenheimer first quoted the line twenty years after the event—appears in Kai Bird and Martin J. Sherwin, *American Prometheus: The Triumph and Tragedy of J. Robert Oppenheimer* (New York: Vintage, 2006).

226   **"It worked":** Frank Oppenheimer told this story in the 1981 documentary *The Day After Trinity*, directed by Jon H. Else.

226   **forty-two scientists:** Connor Nolan et al., "Past and Future Global Transformation of Terrestrial Ecosystems Under Climate Change," *Science* 361, no. 6405 (August 2018): pp. 920–23.

227   **James Lovelock:** His "The Quest for Gaia" was first published in *New Scientist* in 1975, and over the years Lovelock became less and less sanguine. In 2005, he published *Gaia: Medicine for an Ailing Planet*, in 2006 *The Revenge of Gaia*, and in 2009 *The Vanishing Face of Gaia*. He has also advocated geoengineering as a last-ditch effort to stop climate change.

227   **"spaceship earth":** Buckminster Fuller popularized the term, but it appeared originally almost a century before him, in Henry George's 1879 work *Progress and Poverty*—in a passage later summarized by George Orwell in *The Road to Wigan Pier:*

    The world is a raft sailing through space with, potentially, plenty
    of provisions for everybody; the idea that we must all cooperate

and see to it that everyone does his fair share of the work and gets his fair share of the provisions seems so blatantly obvious that one would say that no one could possibly fail to accept it unless he had some corrupt motive for clinging to the present system.

In 1965, Adlai Stevenson managed to give a more poetic treatment, in an address before the United Nations Social and Economic Council in Geneva:

> We travel together, passengers on a little space ship, dependent on its vulnerable reserves of air and soil; all committed for our safety to its security and peace; preserved from annihilation only by the care, the work, and, I will say, the love we give our fragile craft. We cannot maintain it half fortunate, half miserable, half confident, half despairing, half slave—to the ancient enemies of man—half free in a liberation of resources undreamed of until this day. No craft, no crew can travel safely with such vast contradictions. On their resolution depends the survival of us all.

Index